PULMONARY AND RESPIRATORY DISEASES AND DISORDERS

ACUTE LUNG INJURY

EPIDEMIOLOGY, HEALTH EFFECTS AND THERAPEUTIC TREATMENT STRATEGIES

PULMONARY AND RESPIRATORY DISEASES AND DISORDERS

Additional books in this series can be found on Nova's website under the Series tab.

Additional e-books in this series can be found on Nova's website under the e-book tab.

Pulmonary and Respiratory Diseases and Disorders

Acute Lung Injury

Epidemiology, Health Effects and Therapeutic Treatment Strategies

Daniela Mokrá, M.D., Ph.D.
Editor

New York

Copyright © 2014 by Nova Science Publishers, Inc.

All rights reserved. No part of this book may be reproduced, stored in a retrieval system or transmitted in any form or by any means: electronic, electrostatic, magnetic, tape, mechanical photocopying, recording or otherwise without the written permission of the Publisher.

For permission to use material from this book please contact us:
Telephone 631-231-7269; Fax 631-231-8175
Web Site: http://www.novapublishers.com

NOTICE TO THE READER

The Publisher has taken reasonable care in the preparation of this book, but makes no expressed or implied warranty of any kind and assumes no responsibility for any errors or omissions. No liability is assumed for incidental or consequential damages in connection with or arising out of information contained in this book. The Publisher shall not be liable for any special, consequential, or exemplary damages resulting, in whole or in part, from the readers' use of, or reliance upon, this material. Any parts of this book based on government reports are so indicated and copyright is claimed for those parts to the extent applicable to compilations of such works.

Independent verification should be sought for any data, advice or recommendations contained in this book. In addition, no responsibility is assumed by the publisher for any injury and/or damage to persons or property arising from any methods, products, instructions, ideas or otherwise contained in this publication.

This publication is designed to provide accurate and authoritative information with regard to the subject matter covered herein. It is sold with the clear understanding that the Publisher is not engaged in rendering legal or any other professional services. If legal or any other expert assistance is required, the services of a competent person should be sought. FROM A DECLARATION OF PARTICIPANTS JOINTLY ADOPTED BY A COMMITTEE OF THE AMERICAN BAR ASSOCIATION AND A COMMITTEE OF PUBLISHERS.

Additional color graphics may be available in the e-book version of this book.

Library of Congress Cataloging-in-Publication Data

ISBN: 978-1-61470-426-3
Library of Congress Control Number: 2014948643

Published by Nova Science Publishers, Inc. † New York

Contents

Preface		vii
Chapter 1	TRALI Not a Two Hit, But a Multi-Causal Model *Rutger A. Middelburg and Johanna G. van der Bom*	1
Chapter 2	Mechanisms and Prevention of Pulmonary Microvascular Hyperpermeability in Acute Lung Injury *Qiang Zhao, Qin Liu and Jinzhou Zhang*	19
Chapter 3	Pharmacological Therapies for Acute Respiratory Distress Syndrome *Hironobu Hamada*	37
Chapter 4	Novel Trends in Pharmacotherapy of Acute Lung Injury and Acute Respiratory Distress Syndrome *Daniela Mokrá and Juraj Mokry*	45
Editor Contact Information		77
Index		79

PREFACE

In this book, authors review the current knowledge on acute lung injury. Topics discussed include description of different models of etiology of transfusion related acute lung injury (TRALI); mechanisms and prevention of pulmonary microvascular hyperpermeability in acute lung injury; and two reviews on novel therapies of acute lung injury.

Chapter 1 - The etiology of transfusion related acute lung injury (TRALI) is often described as a two hit model. The first hit being for example surgery, infection, hypoxia, trauma, cardio-pulmonary bypass, sepsis or extensive burns. These conditions first activate the pulmonary endothelium and prime neutrophils for activation. The second hit then consists of the transfusion of a blood products containing so called biological response modifiers, which include lipids, peptides, cytokines or antibodies. These substances will activate the already primed neutrophils, leading to the onset of TRALI.

Like almost any disease, TRALI is not caused by two hits but simply multi-causal. For example, a minor surgery combined with a mild infection might have a similar priming effect as a period of cardio-pulmonary bypass. Both would require the subsequent action of a "trigger" (i.e. the transfusion), to cause the onset of TRALI. However, even though the priming already consists of two "hits" (i.e. the combination of two mild risk factors), we still do not call this a three-hit model. This is generally recognized as the multi-causality we typically observe in almost any disease.

In this chapter the author will show how two different models of multi-causality, both of which are commonly used in the description of the etiology of diseases, can be applied to TRALI. First, she will discuss the intuitively appealing threshold model, which shows some similarity with the TRALI susceptibility model previously suggested by Bux and Sachs. Second, she will introduce the more abstract sufficient cause model. Finally, she will show how these two seemingly very different models can both be used to describe the same observed relations between multiple risk factors and TRALI.

Chapter 2 - The ability of the lung to perform gas exchange is made possible in part due to the effective relationship between the alveolar epithelium and the endothelium of the pulmonary microvasculature. Dysfunction of the normal endothelial–epithelial barriers plays a fundamental role in the development of acute lung injury (ALI), a life-threatening syndrome that causes high morbidity and mortality and leads to acute hypoxemic respiratory failure. ALI is characterized by increased endothelial and epithelial permeability, edema, uncontrolled neutrophils migration to the lung, and diffuse alveolar damage. Two leading pathogenic mechanisms of ALI are increased endothelial permeability and reduced alveolar liquid clearance capacity. In most cases, increased endothelial permeability draws more attention since it is the first step of the occurrence and development of ALI. Understanding of the fundamental mechanisms involved in the regulation of endothelial permeability is essential for the development of barrier protective therapeutic strategies. In addition, growing evidence suggests that selected common mechanisms contributing to endothelial barrier protection may be shared by different barrier protective agents, although each agonist triggers a unique pattern of signaling pathways. Understanding of basic barrier protective mechanisms in pulmonary endothelium is essential for the selection of optimal treatment of pulmonary edema of different etiology. Therefore, this review article focuses on the mechanisms and development of lung vascular permeability, and reviews inflammatory cells, cytokines and major intracellular signaling cascades involved in the preservation of the endothelial monolayer barrier.

Chapter 3 - Although many pharmacological therapies, including corticosteroids, neutrophil elastase inhibitors, and anticoagulants, have been evaluated for use in patients with acute respiratory distress syndrome (ARDS), most of these have failed to reduce mortality or improve patient outcomes. However, some therapies have shown beneficial effects for certain subgroups of patients. This review provides a summary of pharmacological therapies used for patients with ARDS.

Chapter 4 - Acute respiratory distress syndrome (ARDS) and its milder form acute lung injury (ALI) occur as a result of various diseases and situations including sepsis, pneumonia, trauma, acute pancreatitis, aspiration of gastric contents, near-drowning etc. ALI/ARDS is characterized by diffuse alveolar damage, alveolar capillary leakage, lung edema, neutrophil-derived inflammation, and surfactant dysfunction. These changes lead to clinical manifestations of decreased lung compliance, severe hypoxemia, and bilateral pulmonary infiltrates. Appropriate treatment involves protective lung ventilation, optimal fluid management, and pharmacological treatment. Despite there are several possibilities of potentially beneficial pharmacotherapy of ALI/ARDS, such as pulmonary vasodilators, corticosteroids, antioxidants, methylxanthines, or exogenous surfactant, the results of experimental and clinical studies are often controversial. This chapter reviews the use of various pharmacological agents and critically evaluates their effects in animal models and in the patients with ALI/ARDS.

In: Acute Lung Injury
Editor: Daniela Mokrá

ISBN: 978-1-61470-426-3
© 2014 Nova Science Publishers, Inc.

Chapter 1

TRALI NOT A TWO HIT, BUT A MULTI-CAUSAL MODEL

Rutger A. Middelburg* and Johanna G. van der Bom
Center for Clinical Transfusion Research, Sanquin Research, Leiden;
Department of Clinical Epidemiology, Leiden University Medical Center,
Leiden, The Netherlands

ABSTRACT

The etiology of transfusion related acute lung injury (TRALI) is often described as a two hit model. The first hit being for example surgery, infection, hypoxia, trauma, cardio-pulmonary bypass, sepsis or extensive burns. These conditions first activate the pulmonary endothelium and prime neutrophils for activation. The second hit then consists of the transfusion of a blood product containing so called biological response modifiers, which include lipids, peptides, cytokines or antibodies. These substances will activate the already primed neutrophils, leading to the onset of TRALI.

Like almost any disease, TRALI is not caused by two hits but simply multi-causal. For example, a minor surgery combined with a mild infection might have a similar priming effect as a period of cardio-pulmonary bypass. Both would require the subsequent action of a

---

* Corresponding author: Rutger A Middelburg, PhD., Center for Clinical Transfusion Research, Sanquin Research, Leiden & Department of Clinical Epidemiology, Leiden University Medical Center, Plesmanlaan 1a, 2333 BZ, Leiden, The Netherlands. Email: R.A.Middelburg@lumc.nl.

"trigger" (i.e. the transfusion), to cause the onset of TRALI. However, even though the priming already consists of two "hits" (i.e. the combination of two mild risk factors), we still do not call this a three-hit model. This is generally recognized as the multi-causality we typically observe in almost any disease.

In this chapter we will show how two different models of multi-causality, both of which are commonly used in the description of the etiology of diseases, can be applied to TRALI. First, we will discuss the intuitively appealing threshold model, which shows some similarity with the TRALI susceptibility model previously suggested by Bux and Sachs. Second, we will introduce the more abstract sufficient cause model. Finally, we will show how these two seemingly very different models can both be used to describe the same observed relations between multiple risk factors and TRALI.

Keywords: Acute lung injury, transfusion, biological response modifiers, etiology, threshold model, sufficient cause model

INTRODUCTION

Transfusion-related acute lung injury (TRALI) is defined as acute lung injury occurring within six hours after a blood transfusion, in the absence of more plausible explanations. [1, 2] Bilateral pulmonary infiltrates and dyspnea should be present in the absence of circulatory overload. [1, 2] TRALI is clinically indistinguishable from acute lung injury or acute respiratory distress syndrome (ARDS) resulting from other causes. This is not surprising, since the final common path in the pathophysiology of TRALI is also identical to that of ARDS. Activation of neutrophils causes damage to the pulmonary vascular lining, which in turn causes leakage of fluid into the alveolar space, under conditions of normal hydrostatic pressure. [3] The key feature of TRALI is the fact that the recipients' neutrophils are activated in response to substances in the transfused blood product. This likely also explains the relatively favorable prognosis of TRALI, which usually resolves spontaneously within 96 hours, [4] probably since the offending substances are cleared. Leukocyte antibodies (i.e. antibodies against human leukocyte antigens or human neutrophil antigens) are generally considered the most important potentially neutrophil activating substances in blood products. [5-7] However, other "biological response modifiers" have also been suggested to be able to activate sufficiently "primed" neutrophils. [8] These biological response modifiers could include cytokines, lipids, peptides and even leukocyte antibodies.

Though leukocyte antibodies were initially postulated to be sufficient to cause TRALI by themselves, [9] it was later realized that this was probably true for only a relatively minor subset of leukocyte antibodies. This limited subset of antibodies could cause TRALI in healthy volunteers and were probably of unusually high affinity and with unusually high neutrophil activating activity. [3] All other transfusion-related mediators, including antibodies with lower intrinsic neutrophil activating activity, would require priming of neutrophils or activation of endothelial cells prior to the blood transfusion. [3, 10] This pathophysiological model was therefore called the "two-event" or "two-hit" model, the first event or hit being the priming of pulmonary neutrophils or activation of pulmonary endothelium by for example surgery, infection, hypoxia, trauma, cardio-pulmonary bypass, sepsis or extensive burns, the second event being a blood transfusion containing biological response modifiers. [10]

This view of the pathogenesis of TRALI suggests there are only two causal components in the etiology of TRALI. Yet, there are a multitude of different causal components necessary to precipitate the onset of any one case of TRALI. As with almost any disease, the etiology of TRALI is, without a doubt, multi-causal in nature. Etiologic processes can not be classified as being four-hit, five-hit, or twenty-hit. First, the number of events depends largely on how we define events and therefore distinguish one event from the other. Second, there are almost always many different ways of causing the same disease, each one potentially consisting of a different number of events. It is therefore thoroughly inaccurate and confusing to refer to the etiology of TRALI as being a two-event phenomenon. This approach doesn't only fail to do justice to the complexity of the pathophysiology of TRALI, as we will show in this chapter, it also hampers etiologic research. By failing to realize the multitude of causes involved in the etiology of TRALI, we obscure the multitude of possible points of intervention for the prevention of TRALI. By lumping together many different patient-related risk factors we lose sight of which patient-related risk factors could be potentially avoided. Therefore, the "two-event model" causes us to miss important opportunities to improve on patient safety. We might end up concluding that we will have to accept a certain incidence of TRALI, simply because we have exhausted our options for the reduction of transfusion-related risk factors. At the same time there could still be an unexploited wealth of patient-related risk factors to intervene on, which we completely fail to recognize, because we over-simplistically lumped them together as "patient predisposition".

THRESHOLD MODELS

An intuitively appealing way of describing multi-causality is by using a threshold model. In a threshold model we assume we can define a single biological system which causes the disease when it becomes (over) activated (i.e. a final common pathway, activated in all cases of the disease). Each risk factor for the disease raises the "activation status" of this system to some extent, until a critical threshold level is reached and clinical disease ensues. Bux and Sachs suggested a simple threshold model for the etiology of TRALI. [3] More elaborate threshold models, that more fully reflect the multi-causality of disease, had previously been suggested for the etiology of venous thrombosis. [11] Here we will first review some of the strengths and limitations of the Bux-Sachs threshold model and consider the more general threshold model of multi-causality. Then we will apply the multi-causal threshold model to the specific case of the etiology of TRALI.

Bux-Sachs Threshold Model of TRALI

Like any threshold model, the Bux-Sachs threshold model assumes there is a single system involved in all cases of TRALI. Not surprisingly this system is defined as the combination of pulmonary neutrophils and endothelium. If the combination of these gets activated beyond a certain level, the vasculature will become permeable to fluids at normal hydrostatic pressure, fluid leakage into the alveolar space will occur and TRALI will ensue. Bux and Sachs hypothesized there are two distinct thresholds one for mild and one for severe TRALI. This might be a useful extension of the threshold model, however it has no implications for the conceptual understanding of either the threshold model or multi-causality of TRALI and we will therefore ignore the double threshold for the rest of this chapter.

The Bux-Sachs model consists of three axes (see figure 1 for an adaptation of the original model). There's a continuous axis for the state of activation of the pulmonary neutrophil/endothelial system (the ordinate) and there are two parallel axes (the abscissae). The abscissa on the top of the figure represents patient predisposition, while the abscissa at the bottom of the figure represents (in reverse direction) the activating potential of a blood transfusion. This choice of axes seems to stem directly from the two-event model. It assumes the transfusion is one event and all other risk factors are lumped together as the other of a total of two events. All these lumped together pre-transfusion risk factors are then referred to as "individual predisposition".

The model's greatest strength is that it formalizes the concept that more vulnerable (i.e. predisposed) patients would require less activation of their pulmonary neutrophil/endothelial system by a transfusion to develop TRALI. Conversely, healthy individuals would require very strong activation by the transfusion. In other words, on the left side of the graph transfusion-related activation is high (much white area), so very little patient-related activation is needed (little grey area), while on the right side of the graph transfusion-related activation is low (little white area), so only heavily predisposed patients (much grey area) will develop TRALI.

The main limitation of this model is that it essentially carries similar limitations as the two-event model. There is no distinction between different pre-transfusion risk factors, while this distinction could be crucial to identify potentially avoidable risk factors. This threshold model therefore fails to do full justice to the multi-causality of TRALI, because it was based on the two-event model and as a consequence lumps all patient-related risk factors together as one (i.e. patient predisposition) and offsets this against the transfusion's neutrophil/endothelium activating potential.

A More General Threshold Model of Multi-Causality

Like the Bux-Sachs threshold model for TRALI, more general threshold models of multi-causality must also put activation of the disease causing system on the ordinate. However, by putting time on the abscissa, they also create the possibility of representing factors that only temporarily increase the activation (i.e. over time an increased level of activation can decrease again). Further, by not putting some risk factors on one of the axes, these models allow all risk factors to be treated similarly. Each risk factor, at the time of occurrence, gives an increase in activation status. All these increases will add up until the threshold is reached and disease ensues (figure 2). This type of model has previously been used by Rosendaal to illustrate the multi-causality of venous thrombosis. [11] Here we illustrate the Rosendaal model based on four simple hypothetical scenarios involving the example of TRALI (figure 2).

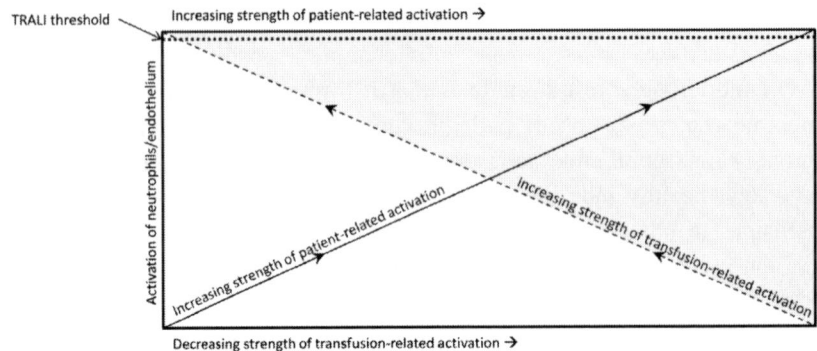

Figure 1. Bux-Sachs threshold model. [3] TRALI occurs when activation of neutrophils and endothelium crosses the threshold. From left to right the contribution of patient-related risk factors (grey area) increases. Since the total amount of neutrophil and endothelial activation needed to cross the threshold is always the same, this automatically means the (minimal required) contribution of transfusion-related risk factors (white area) must decrease. Therefore, healthy individuals (far left) need extremely strong neutrophil and endothelial activation potential from the transfusion, while heavily predisposed patients (far right) need only a very mildly stimulating transfusion to develop a case of TRALI.

Rosendaal Threshold Model Applied to TRALI

To illustrate the application of the Rosendaal threshold model for the description of TRALI etiology, we consider different possible scenarios for a hypothetical patient (figure 2). This patient experiences three distinct events potentially predisposing for TRALI:

1. Severe pneumonia, requiring lengthy intensive care unit admission and mechanical ventilation, raises the activation status of pulmonary neutrophils and endothelium substantially and for a long period.
2. Uncomplicated cardiac surgery, requiring some time on cardio-pulmonary bypass, causes an intermediate increase in the activation status of pulmonary neutrophils and endothelium for a short period.
3. Hematologic malignancy causes a slight increase in the activation status of pulmonary neutrophils and endothelium for an intermediate period.

This patient also receives two blood transfusions that each increase the activation status of pulmonary neutrophils and endothelium for a very short time with transfusion X giving half the activation of transfusion Y. Other transfusions might also have been given, but these did not increase the activation of pulmonary neutrophils and endothelium and are therefore not depicted.

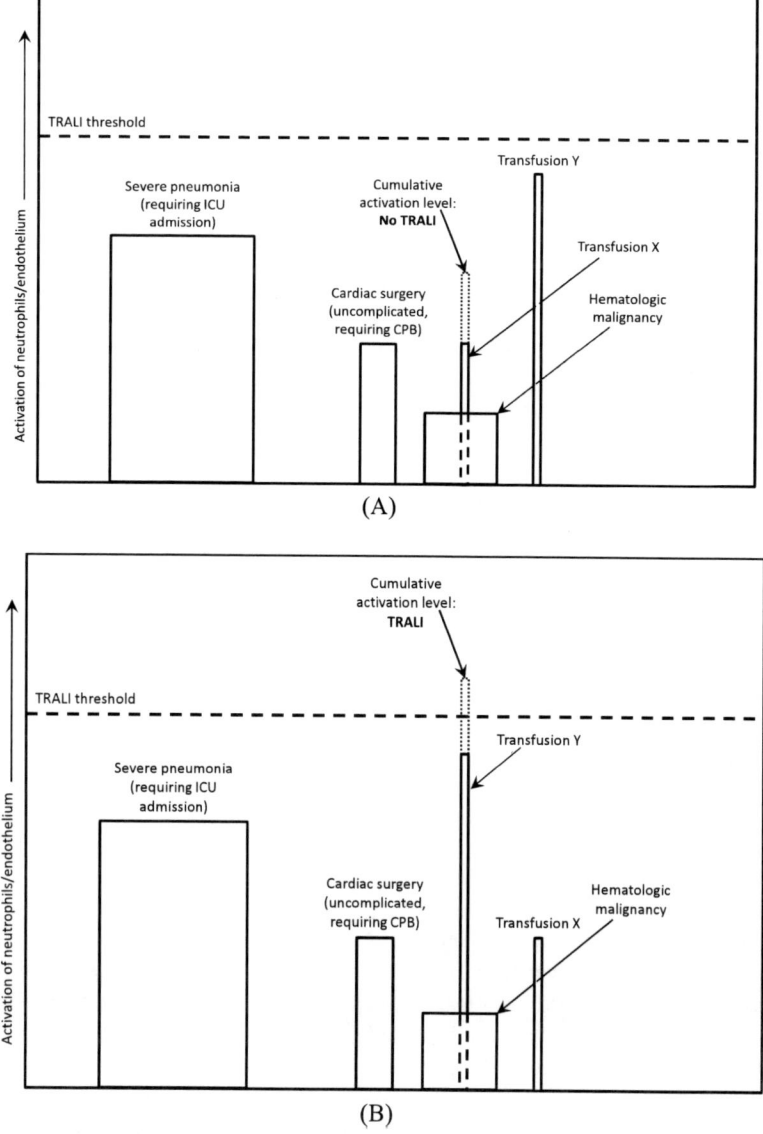

Figure 2. (Continued).

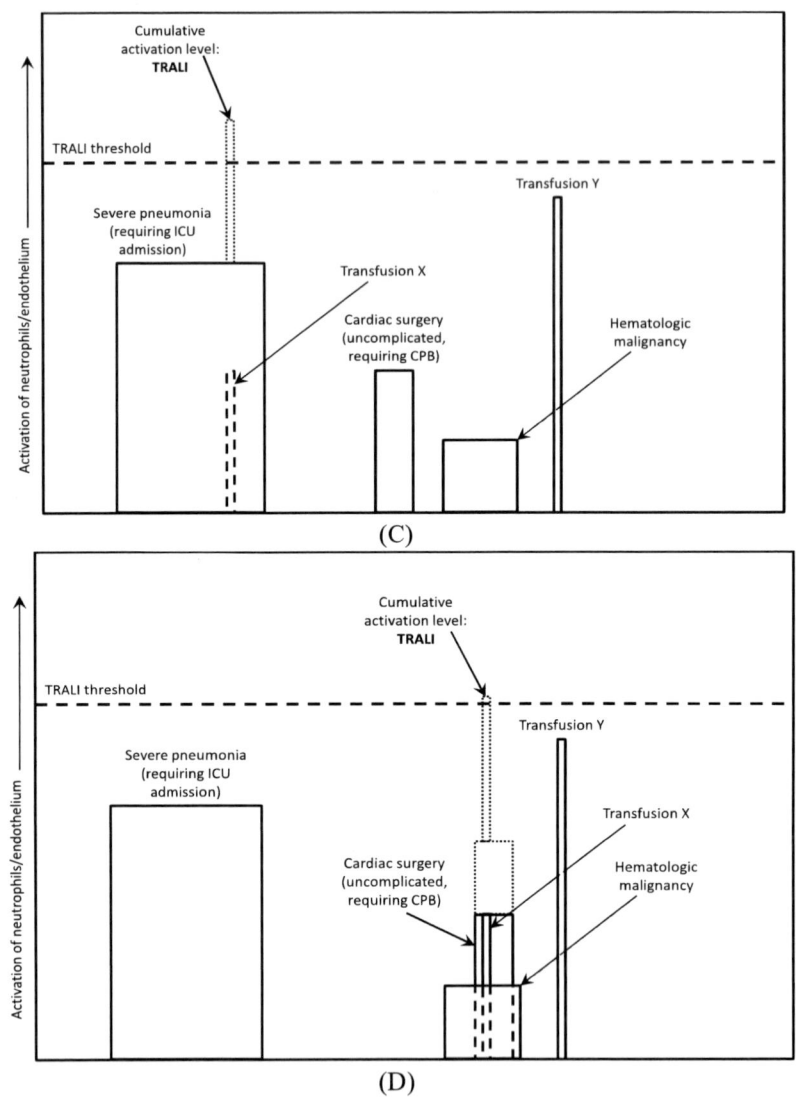

Figure 2. Rosendaal threshold model of multi-causality,[11] applied to TRALI. The level and duration of activation of pulmonary neutrophils and endothelium by different events is indicated by solid lined blocks. Dashed lines (long dashes) indicate two events coinciding (i.e. one passing "behind" the other). The cumulative activation is determined by stacking these boxes on top of each other. The resulting cumulative level is indicated by the finely dashed lines. If the cumulative level passes the TRALI threshold (dashed horizontal line), TRALI ensues. See text for a detailed description of all depicted scenarios.

There are four different hypothetical scenarios depicted:

A. Transfusion X is given during the period in which the activation of pulmonary neutrophils and endothelium was raised by hematologic malignancy. Since both transfusion X and hematologic malignancy are only mild activators of pulmonary neutrophils and endothelium, the cumulative activation level does not pass the threshold level and there is no TRALI. Transfusion Y is given during a period without predisposing factors and, although a strong activator, transfusion Y is by itself not sufficient to cause TRALI (i.e. the threshold is again not passed).
B. Transfusions X and Y are switched, causing transfusion Y to be given during hematologic malignancy, instead of during a period without predisposing factors. Because transfusion Y is twice as strong an activator as transfusion X, the combination with only the minor predisposition caused by hematologic malignancy is enough to cause TRALI.
C. Transfusion X is given during the severe pneumonia requiring intensive care unit admission and mechanical ventilation. Since this is a very strong predisposing factor, the minor activation of pulmonary neutrophils and endothelium caused by transfusion X is enough to raise the cumulative level above the threshold, causing TRALI.
D. Cardiac surgery is performed during the period of hematologic malignancy. This raises the predisposition level to almost the same level as seen during severe pneumonia requiring intensive care unit admission and mechanical ventilation. Due to the strong predisposition, the relatively mild activation of pulmonary neutrophils and endothelium by transfusion X is again enough to cause TRALI.

Other scenarios are also possible and some of these will be discussed below as examples of the strengths and limitations of the Rosendaal threshold model. In general, since the Rosendaal threshold model depicts time and there are an infinite number of ways of timing the different events relative to one another, there is an infinite number of scenarios possible. However, only a limited set will be different from one another in any relevant way (i.e. in a way that would affect the causal mechanism responsible for the onset, or lack thereof, of TRALI).

Implications of the Rosendaal Threshold Model

As can be seen from figure 2, the timing of different predisposing events relative to transfusions that cause different levels of activation of pulmonary neutrophils and endothelium is crucial for the etiology of TRALI. If cardiac surgery had been performed during the severe pneumonia, acute lung injury or acute respiratory distress syndrome might even have ensued without the involvement of any transfusions. This could be a reason why a cardiothoracic surgeon might for example opt to postpone surgery until the pneumonia resolved. To prevent TRALI one could theoretically opt to only give transfusions with little to no activating effect on pulmonary neutrophils and endothelium. However, this could prove logistically impossible or prohibitively expensive. In this case a reasonable second option could be to give high risk transfusions to low risk patients only and give only low risk transfusions to high risk patients. For this to be a realistic alternative, we need to be able to determine which patients are at the highest predisposition level. This is determined multi-causally and can therefore not be appropriately studied using a two-event model which lumps all predisposition together. Moreover, we might need to decide to postpone cardiac surgery during hematologic malignancy, not because this combination is dangerous in itself, but because the combination is predisposing enough to make even intermediate risk transfusions dangerous (figure 2D). Unless we can guarantee transfusions carrying an extremely low TRALI risk, this could be a relevant consideration. Finally, the level of activation of pulmonary neutrophils and endothelium typically associated with the majority of transfusions could be in the order of magnitude depicted for transfusion X rather than that depicted for transfusion Y. In this case, we might need to study patients receiving transfusions while undergoing cardiac surgery during hematologic malignancy, and compare these to patients receiving transfusions while undergoing cardiac surgery without hematologic malignancy, to even be able to find out that hematologic malignancy is indeed a predisposing condition for TRALI. We need multi-causal models, to be able to adequately represent this scenario.

THE SUFFICIENT CAUSE MODEL

A more abstract way of representing the multi-causality of disease was suggested in 1976 by Kenneth J. Rothman. [12] The sufficient cause model distinguishes three different types of causes: necessary, component, and sufficient causes.

A component cause is any cause that contributes to the total causal mechanism. In the case of TRALI this could be any of the examples given in figure 2: severe pneumonia, uncomplicated cardiac surgery, hematologic malignancy, or transfusion.

A sufficient cause is any combination of different component causes which is sufficient to cause disease. Figure 3 represents all possible sufficient causes for the example from figure 2. For example in figure 2D the sufficient cause is the combination of uncomplicated cardiac surgery, hematologic malignancy, and transfusion X (figure 3F). Since we will never know all contributing factors for a disease, we usually also include an "unknown" component cause (i.e. U in figure 3), which can be further divided into several known component causes by further research. The model assumes a person is "waiting" for enough component causes to accumulate to complete a sufficient cause of disease. This can be represented graphically as a pie accumulating pieces until the entire pie is filled, as shown in figure 3. The model is therefore also often popularly referred to as the *"causal pie model"* and people can be thought of as *"walking around with multiple incomplete pies"*.

A necessary cause is any component cause that is necessary in all cases of the disease. In the case of TRALI a transfusion is by definition always necessary and could therefore be considered a necessary cause. However, it is usually more informative to consider the differences between seemingly identical component causes, in our example the difference between transfusions X and Y.

Let us suppose transfusion X contains relatively weak biologic response modifiers and transfusion Y contains strong leukocyte antibodies. In our example we should distinguish these and consider them as different types of component causes. Transfusion Y could complete sufficient causes containing any of the other three component causes by themselves or in combination (figure 3A-D), while transfusion X can only complete sufficient causes containing either severe pneumonia (figure 2C, 3E), or a combination of the other component causes (figure 2D, 3F).

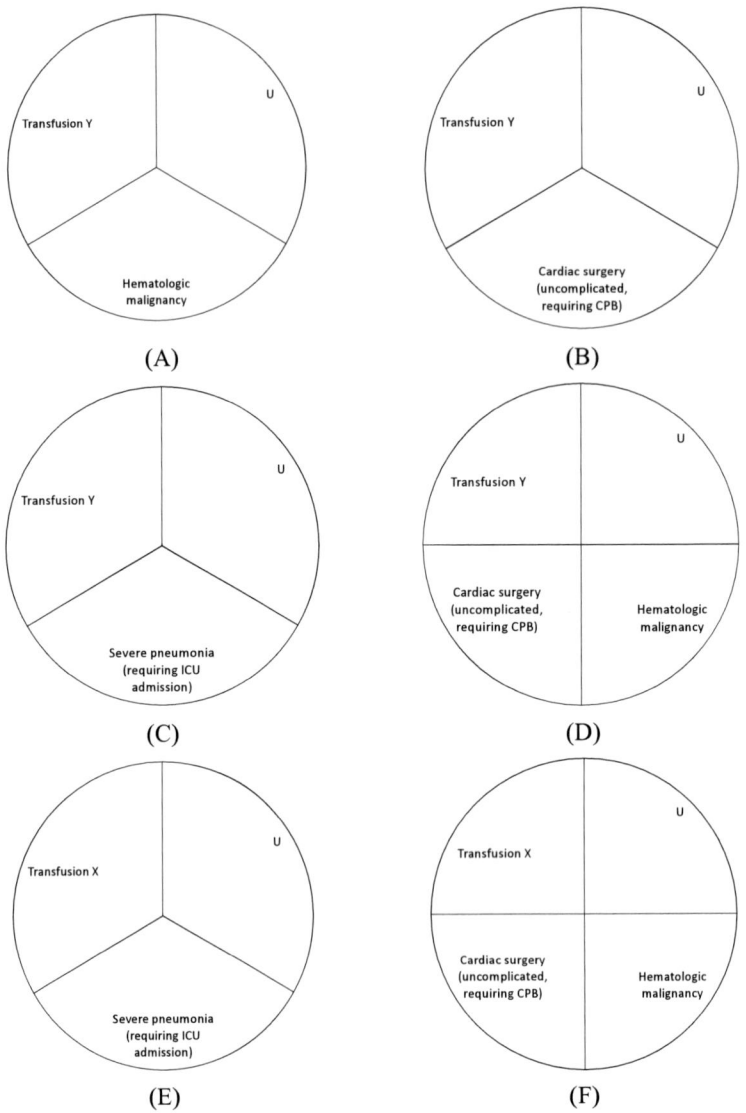

Figure 3. Continued on next page.

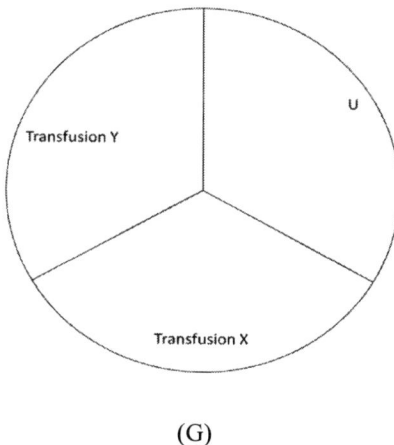

(G)

Figure 3. Rothman's sufficient cause model applied to the TRALI example from figure 2. Each "pie" represents a sufficient cause of TRALI. Each "piece of a pie" represents a component cause. If all component causes are present that particular sufficient cause is completed and TRALI ensues. Each person can have a multitude of uncompleted sufficient causes, but only the first one to be completed is considered the cause of that case of TRALI. A single component cause can be present in multiple sufficient causes (e.g. transfusion Y in A through D and G). "U" represents all unknown other component causes necessary to complete a given sufficient cause. Each "U" can represent a different set of unknown component causes, because different additional component causes might be required depending on which other component causes are present.

Multiple Potential Sufficient Causes

Crucial for the sufficient causes model is the idea that one person can live with any number of partly completed sufficient causes. The first cause that happens to be completed is the one that actually causes the disease, making all other potential (partly completed) causes irrelevant in the instant disease occurs. For example, if in figures 2C or 2D transfusion Y would have been given shortly after transfusion X, this would have again caused the level of neutrophil/endothelial activation to cross the threshold. Although this could theoretically be considered a sufficient cause of TRALI (figure 3C or D), transfusion Y would not be considered a cause of TRALI in this scenario, since there was a different sufficient cause that was completed previously (figure 3E or F).

Limitations of the Sufficient Cause Model

The sufficient cause model knows some limitations which can lead to a confusion of different sufficient causes, especially if multiple sufficient causes are completed simultaneously. For example, there is no problem if a person has severe pneumonia. This person has both sufficient causes represented in figures 3C and E partly filled and which of the two is completed is decided by whether this person receives transfusion X or Y. As discussed above, if this person receives both transfusions, the first sufficient cause to be completed is considered the cause of this case of TRALI. Likewise, a person with hematologic malignancy has the sufficient causes in figures 3A, D and F partly filled. If this person then receives transfusion Y the cause in figure 3A will be completed. Anything that happens afterwards, even if it theoretically completed the sufficient causes represented in figures 3D and F is then irrelevant. However, if this person with hematologic malignancy first has cardiac surgery and gets the transfusion later, the causes in figures 3D and F would apply instead of the one from figure 3A. Which of these two would then be completed would again depend on which transfusion is given. If transfusion Y is subsequently given, we might say the cause in figure 3D is completed. However, it could be validly argued that the causes in figure 3A and B are also completed.

The sufficient cause model solves this problem by requiring the simplest sufficient cause possible. In our example this would not be the one represented by figure 3D. Given the combination of hematologic malignancy and transfusion Y, cardiac surgery would no longer be required to cause TRALI. Likewise, given the combination of cardiac surgery and transfusion Y, hematologic malignancy would no longer be required either. Therefore, the sufficient cause in figure 3D would be considered over-determined, and therefore not a separate sufficient cause distinct from those in figures 3A and B. However, this would mean the sufficient causes in figures 3A and B would be completed simultaneously. The model does not specify which sufficient cause to "blame" for this case of TRALI.

Also note, that in this situation transfusion X still unambiguously completes the sufficient cause represented in figure 3F. This would mean that for the same patient (i.e. undergoing cardiac surgery while having a hematologic malignancy) transfusion X would complete a single sufficient cause, while transfusion Y would complete two sufficient causes simultaneously, leaving us with an impossible choice of which of the two predisposing factors contributed causally. At the same time the clinical results of transfusions X and Y would likely be indistinguishable (i.e. both result in TRALI).

Another limitation relates to duration of effect of a component cause. As can be seen from figure 2, the combination of transfusions X and Y is enough to cross the threshold, making figure 3G a sufficient cause of TRALI. This however, requires the transfusions to be given closely together which can not be readily seen from the "*causal pie*" representing this mechanism. This suggests pieces of pie can also disappear again over time, but there is no formal way of representing this in the model.

DISCUSSION

TRALI Not a Two Hit, but a Multi-Causal Model

TRALI, like any disease, is multi-causal. The number of identified causes depends mostly on the amount of research spend on identifying those causes. Considering the etiology of TRALI a "two-event" mechanism is therefore overly simplistic. It lumps all patient-related risk factors together as one predisposing "event" and adds all transfusion-related risk factors as a single transfusion "event". Therefore, the two-event model tends to discourage distinguishing further among the different patient-related, or different transfusion-related risk factors. Making these distinctions can be crucial for the prevention of TRALI.

Threshold versus Sufficient Cause Model

We have discussed two different commonly used models for multi-causality of disease: the intuitively appealing threshold model and the more abstract sufficient cause model. We have shown some of the limitations of the sufficient cause model, which relate mostly to timing of different component causes. Why then, would we not simply use only the threshold model? It solves these problems by incorporating time, is intuitively appealing, is capable of representing any number of causal components that act simultaneously, and reflects differences in effect sizes of different causal components. This is another advantage over the sufficient cause model, in which there is no difference in importance of different component causes. In the sufficient cause model it is assumed all components are necessary to complete a given sufficient cause and therefore all component causes are equally important. At the population level a given component cause can be more important than another because it's a component in more sufficient causes (i.e. has contributed to the completion of the sufficient cause in more patients). At the individual level, however, only a single sufficient cause can be completed and all component causes are considered to have contributed equally. Consider, for example, the causal mechanism represented in figure 2B, which corresponds to the sufficient cause represented in figure 3A. From figure 2B we would say transfusion Y had a bigger contribution to this case of TRALI than the hematologic malignancy. Conversely, from figure 3A, we conclude that if either transfusion Y or the hematologic malignancy had not occurred this would have prevented this case TRALI and since the two component causes are both necessary, they are equally important.

However, the threshold model also has one big disadvantage. As noted above, the different effect sizes represented in figure 2 seem to imply that transfusions X and Y together could cause TRALI. This possibility is reflected by the sufficient cause represented in figure 3G. However, this is probably not, or not always, true. Let us assume both transfusions X and Y contain very high titers of high affinity neutrophil antibodies. The antibodies in transfusion Y happen to be more activating than those in transfusion X, although they recognize adjacent epitopes. In this scenario the activation would not add up. Rather, adding transfusion X to transfusion Y would probably lower the activation normally caused by transfusion Y. The weakly activating antibodies in transfusion X would compete for binding with the more strongly activating antibodies in transfusion Y. There is no way of representing this possibility in the threshold model. In the sufficient cause model, however, this is solved by deleting the sufficient cause represented by figure 3G from the list of possible sufficient causes.

Further, in the sufficient cause model the "U" representing, as yet, unkown other component causes does not need to be specified. It is important to realize it can be a different combination of unkown component causes for each different sufficient cause represented. For example, in figure 3A "U" could include a genetic variant that makes the pulmonary endothelium more vulnerable to the effects of some chemotherapeutic agents commonly used in hematologic malignancies. In figure 3B this variant would be irrelevant, because the patient is not receiving the chemotherapeutic agent in question. Instead the "U" in figure 3B might include the (lack of) use of steroids during surgery. Steroids could change the sensitivity to activation by cardio-pulmonary bypass, while they might be both absent and irrelevant for a patient with hematologic malignancy. These differences do not need to be specified and therefore also do not need to be known. It is therefore very easy to represent the full causal mechanism in the sufficient cause model, even if this mechanism is not completely known.

In the threshold model all we could do to reflect unknown causal components is raise the baseline activation status of the pulmonary neutrophils/endothelium.This raised level could change over time, but can not be modified dependent on the other causal components acting at that time. Therefore, the sufficient cause model in some situations offers more flexible possibilities to completely describe all possible ways of causing disease.

CONCLUSION

Both the threshold model and the sufficient cause model have weaknesses. However, they are both capable of describing in detail all that is currently known about the multi-causal nature of TRALI and both allow for adding new knowledge. Full appreciation of the multi-causal nature of TRALI, enables the identification of possibilities for preventive interventions. We should refrain from using a needlessly limiting description like the "two-event" or "two-hit" model for TRALI, or any other disease, because it shuts possible new avenues of investigation.

REFERENCES

[1] Goldman, M; Webert, KE; Arnold, DM; Freedman, J; Hannon, J; Blajchman, MA. Proceedings of a consensus conference: towards an understanding of TRALI. *Transfus Med Rev*, 2005, 19(1), 2-31.

[2] Kleinman, S; Caulfield, T; Chan, P; Davenport, R; McFarland, J; McPhedran, S; Meade, M; Morrison, D; Pinsent, T; Robillard, P; Slinger, P. Toward an understanding of transfusion-related acute lung injury: statement of a consensus panel. *Transfusion*, 2004, 44(12), 1774-1789.

[3] Bux, J; Sachs, UJ. The pathogenesis of transfusion-related acute lung injury (TRALI). *Br J Haematol*, 2007, 136(6), 788-799.

[4] Popovsky, MA; Moore, SB. Diagnostic and pathogenetic considerations in transfusion-related acute lung injury. *Transfusion*, 1985, 25(6), 573-577.
[5] Kopko, PM. Review: transfusion-related acute lung injury: pathophysiology, laboratory investigation, and donor management. *Immunohematol*, 2004, 20(2),103-111.
[6] Bux, J. Transfusion-related acute lung injury (TRALI): a serious adverse event of blood transfusion. *Vox Sang*, 2005, 89(1), 1-10.
[7] Middelburg, RA; van Stein, D; Briet, E;, van der Bom, JG. The role of donor antibodies in the pathogenesis of transfusion-related acute lung injury: a systematic review. *Transfusion*, 2008, 48(10), 2167-2176.
[8] Silliman, CC; Boshkov, LK; Mehdizadehkashi, Z; Elzi, DJ; Dickey, WO; Podlosky, L; Clarke, G; Ambruso, DR. Transfusion-related acute lung injury: epidemiology and a prospective analysis of etiologic factors. *Blood*, 2003, 101(2), 454-462.
[9] Popovsky, MA; Abel, MD; Moore, SB. Transfusion-related acute lung injury associated with passive transfer of antileukocyte antibodies. *Am Rev Respir Dis*, 1983, 128(1), 185-189.
[10] Silliman, CC. The two-event model of transfusion-related acute lung injury. *Crit Care Med*, 2006, 34(5 Suppl), S124-S131.
[11] Rosendaal, FR. Venous thrombosis: a multicausal disease. *Lancet*, 1999, 353(9159), 1167-1173.
[12] Rothman, KJ. Causes. *Am J Epidemiol*, 1976, 104(6), 587-592.

In: Acute Lung Injury
Editor: Daniela Mokrá

ISBN: 978-1-61470-426-3
© 2014 Nova Science Publishers, Inc.

Chapter 2

MECHANISMS AND PREVENTION OF PULMONARY MICROVASCULAR HYPERPERMEABILITY IN ACUTE LUNG INJURY

Qiang Zhao[1], Qin Liu[2] and Jinzhou Zhang[1]*
[1]Department of Cardiovascular Surgery, Xijing Hospital, Fourth Military Medical University, Xi'an, China;
[2]Mechanical Engineering, University of Texas at San Antonio, San Antonio, TX, US

ABSTRACT

The ability of the lung to perform gas exchange is made possible in part due to the effective relationship between the alveolar epithelium and the endothelium of the pulmonary microvasculature. Dysfunction of the normal endothelial–epithelial barriers plays a fundamental role in the development of acute lung injury (ALI), a life-threatening syndrome that causes high morbidity and mortality and leads to acute hypoxemic respiratory failure. ALI is characterized by increased endothelial and epithelial permeability, edema, uncontrolled neutrophils migration to the lung, and diffuse alveolar damage. Two leading pathogenic mechanisms

---

* Corresponding Author address: Jinzhou Zhang, Department of Cardiovascular Surgery, Xijing Hospital, Fourth Military Medical University, Xi'an 710032, China. Email: jinzhouzhang 2006@yahoo.com.

of ALI are increased endothelial permeability and reduced alveolar liquid clearance capacity. In most cases, increased endothelial permeability draws more attention since it is the first step of the occurrence and development of ALI. Understanding of the fundamental mechanisms involved in the regulation of endothelial permeability is essential for the development of barrier protective therapeutic strategies. In addition, growing evidence suggests that selected common mechanisms contributing to endothelial barrier protection may be shared by different barrier protective agents, although each agonist triggers a unique pattern of signaling pathways. Understanding of basic barrier protective mechanisms in pulmonary endothelium is essential for the selection of optimal treatment of pulmonary edema of different etiology. Therefore, this review article focuses on the mechanisms and development of lung vascular permeability, and reviews inflammatory cells, cytokines and major intracellular signaling cascades involved in the preservation of the endothelial monolayer barrier.

Keywords: Acute lung injury, respiratory failure, microvascular hyper-permeability, endothelium, pulmonary edema

INTRODUCTION

The completeness and effectiveness of alveolar epithelium and pulmonary microvasculature endothelium cells (PMVECs) are fundamental for gas exchange, permeability and other functions. Acute lung injury (ALI) and acute respiratory distress syndrome (ARDS) present the syndromes that are caused by alveolar-endothelial barrier impairment and result in hyperpermeability and respiratory failure. Though great progress of epidemiology, pathogenesis, and management in ALI or ARDS have been made over the past few decades, the mechanisms underlying alveolar-endothelial barrier impairment need further investigation since the mortality of ALI/ARDS still ranges from 24% to 60% [1-3]. PMVECs, monocytes, macrophages and myeloid derived cells are the keys of the innate immune response in mediating the host response to ALI [4]. However, mechanisms of ALI are far more complex than have been currently studied. This article focuses on the mechanism and development of lung vascular hyperpermeability in ALI, and reviews inflammatory cells, cytokines and major intracellular signaling cascades that involved in the preservation of the endothelial monolayer barrier.

Pulmonary Microvascular Hyperpermeability in Acute Lung Injury

The alveolar-capillary barrier is the key tissue for pulmonary gas exchange. Two predominant cell types are separated from alveolar epithelial cells: one is type I cells, constituting 20% of the total epithelial cells but accounting for 80% of alveolar surface, and allowing for diffusion and exchange of gas, are susceptible to the injury and cell death; the other is the more resistant type II cells that can proliferate and differentiate into type I cells, which is important in re–epithelialization of epithelial barrier. These cells also produce surfactant and regulate fluid balance across the epithelium. Both type I and type II cells play roles in host defense and immunity [5-7]. PMVECs are spread along the inner layer of microvessels. Dysfunction of PMVECs results in the leakage of protein-rich fluid and neutrophils that are recruited into the pulmonary interstitium and alveolar spaces, and causes alveolar edema [8]. Thus PMVECs is the key component for the integrity of the barrier.

Permeability regulation of PMVECs during ALI/ARDS is the result of a complex interaction among many soluble factors and inflammatory cells. Some can aggravate injury, e.g., high mobility group box protein box1 (HMGB1), NF-κB related molecules, RhoA, neutrophils, and the others are protective, such as alveolar macrophages [9-12]. The barrier dysfunction is of critical importance for the development of ALI [13, 14]. Fluid transport is also regulated by the alveolar epithelium [15]. Following various stimuli, pulmonary epithelial-endothelial barrier becomes dysfunction; this will lead to uncontrolled hyperpermeability, followed by complex cascades, such as the infiltration of macrophage and neutrophil, the activation of cytokines and reactive oxygen species, finally the leakage of protein-rich fluid into the interstitial, and respiratory dysfunction [6, 16].

The Mechanisms of Pulmonary Microvascular Hyperpermeability in Acute Lung Injury

The permeability of the lung is more liable on pulmonary microvascular filtering function of the alveolar–capillary barrier, which is based on integrity and efficiency of the microvascular endothelium, interstitium and alveolar epithelium [17]. The function of the barrier includes at least two parts, the lymphoma flow and alveolar liquid. They form the aqueous subphase to protect the layer of surfactant phospholipids and proteins that lining the alveolar epithelium [13, 18]. The filtration also regulates trans-vascular molecular flow such as water, proteins, and small solutes especially the oxygen carbohydrates. Under physiological conditions, liquid in alveolar is cleared by the pressure gradient that directs the fluid formed at the venules towards the lymphatic, therefore promoting interstitial fluid clearance [13, 19]. Anything causes the edema in septal and alveolar, such as elevated post-capillary microvascular pressure, breaks the balance of filtration and eventually impairs oxygen diffusion [13, 20].

Macrophages

Alveolar macrophages play roles in recognizing pathogens, initiating host defense via protective inflammation, and clearing pathogens from the alveolar. Alveolar macrophages initiate the immune response form the first line of defense in pathogen elimination, and in later stages they promote the clearance of apoptotic neutrophils and excess mononuclear, allowing inflammation resolution and tissue repair [11, 12]. Macrophages response to multiple cytokines such as tumor necrosis factor (TNF)-α, NF-κB pathway, stimulate epithelial cells and tissue-resident macrophages, mediate the recruitment of neutrophils, exudate macrophages and lymphocytes to the site of injury, and ultimately clear pathogens.

Neutrophils

Despite its important role in pathogen containment, excessive alveolar neutrophil recruitment has been associated with the injury to the alveolar-capillary barrier in ALI [21]. Neutrophils recruit into the alveolar compartment, release proteases or neutrophil extracellular traps that composing of chromatin and antimicrobial agents, and cause endothelial or epithelial injury [22, 23]. Neutrophils have been established as a major pathway in experimental models of sepsis, pneumonia, ischemia-reperfusion, trauma and shock, and transfusion-associated lung injury [14, 24]. Neutrophil adhesion and motility is regulated by Hax1 through RhoA [25].

Platelets

Platelets express interleukin (IL)-1, thromboxane A_2, Toll-like receptor(TLR)-2, TLR-4 etc., which are cellular components of inflammation and coagulation cross talk to fight pathogens of the lung by complex pathways [26, 27]. However, if the balance is broken, the reaction of these molecules may be a disaster to the vulnerable endothelial-epithelial barrier of the lung. General inflammatory molecules and signaling pathway were recently shown to contribute to lung injury. TLR-4 activates nuclear factor-kappa B (NF-κB) and MAPK pathways to regulate the expression of proinflammatory COX-2 in renal medullary collecting duct cells [28]. NF-κB is also known as a regulator of mass inflammatory processes. Reports have shown that NF-κB plays an important role in the pathogenesis of lung diseases [26, 29]. NF-κB is required for maximal transcription of numerous cytokines including tumor necrosis factor-α, IL-1β, and IL-6 [30].

Cell Organelles

Cell organelles can be dysfunction in ALI, and subsequently lead to the impaired Na^+-K^+-ATPase activity and lung edema. Elevated CO_2 levels, metabolites of NO, and oxidant can induce the dysfunction of mitochondrial, which functions as the repair of the alveolar epithelial barrier [31].

Small G Protein

Small GTPases are considered as "molecular switches", whose cycling between active and inactive forms is regulated stringently by cellular factors. Small GTPase are composed mainly of Ras homolog member A (RhoA), Ras-related C3 botulinumtoxin substrate1 (Rac1) and cell division control protein 42 homolog (Cdc42). However, even these three best-characterized members, with complex functions on vascular permeability, still have not been clarified. Many researchers have reported correlations between altered RhoA, Rac1 and/or cdc42 levels and endothelial cell permeability, but rare on GJs and in ALI. Rho GTPases and connexin may converge to regulate GJs' permeability by either affecting the stability of connexin or by modulating their interactions. Rac1 plays a central role in cell migration, adhesion, and permeability [32]. Rac1 stabilizes endothelial junctions and counteracts the effects of Rho [33,

34]. High expression of Rac1 up-regulate Cx40 [3]. Activation of Rac1 is regarded as a suitable approach to amend barrier functions in sepsis. It has been proven that the activation of Rac1 takes protective effect in both microvascular and macrovascular endothelium [35-37]. Cdc42 is known to promote the formation of actin-rich and fingerlike membrane extensions that regulating cell polarity and protecting endothelial barrier [38, 39]. Activation of Rac1/cdc42 has been found to stabilize the endothelial barrier by regulating the interactions between α-catenin and cadherins [40]. However, cdc42 plays a controversial role in the assembly and maintenance of epithelial junctions [37].

Rho-GTPases is a subgroup of Ras superfamily of 20–30 kD GTP-binding proteins that widely regulate cellular functions [41]. Studies illustrated the activities of RhoA, Rac1 and Cdc42 are involved in cell molecular and physiological functions such as cell adherens ability, motility, migration and endothelial barrier integrity [42-44]. RhoA, Rac1 and Cdc42 regulate separate signal transduction pathways that linking plasma membrane receptors to the assembly of distinct filamentous actin structures, and modulate cell-cell junctions mainly by controlling the assembly and contractility of the actin cytoskeleton [37, 45]. There is clear evidence that connexin functional impairing and intercellular communication disorder are demonstrated by elevated intracellular Ca^{2+} in ALI [46]. The Rho family of small GTPases is known as a major regulator of cellular junctions and of actin cytoskeleton [35]. Inhibition of RhoA has proven to induce the reduction in human umbilical vein endothelial permeability [40].

Rho kinase-mediated myosin light chain phosphatase (MLCP) leads to increased MLC phosphorylation and actin–myosin contractility, endothelial cells contraction, intercellular junction disruption, and barrier hyperpermeability [47]. Contractility of the actin cytoskeleton in PMVECs increases following the activation of MLCK and MLCP, and consequently, intracellular tethering force decreases and pulmonary microvascular permeability increases [35]. Rho GTPase/Rho GTPase kinase (Rho/ROCK) and mitogen-activated protein kinase (MAPK) signaling pathway are involved in these processes. RhoA activation attenuates myosin phosphorylation via Rho GTPase kinase (ROCK), subsequently leads to the upregulation of myosin light chain (MLC) phosphorylation and cytoplasmic actin–myosin cross linking, and finally results in polymerization of the actin microfilament [48, 49]. Myosin light chain kinase (MLCK) takes an important role in cellular reaction in ALI. Phosphorylated MLCK leads to cytoskeletal reorganization and endothelial cells contraction based on actomyosin. Therefore, the balance between peripheral adhesive and concentric forces of endothelial cells is broken, which causes barrier impairing and pulmonary vascular hyperpermeability [13] (Figure 1).

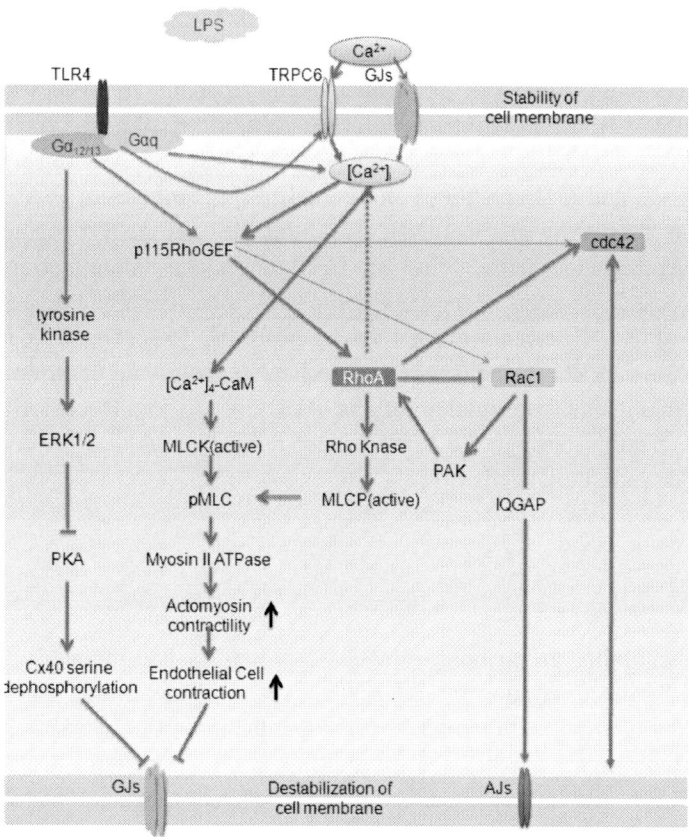

Figure 1. Signal transduction cascade of LPS-induced interference to GJs and cell membrane stability in PMVECS. The barrier destabilizer RhoA-related signaling pathway is targeted by LPS, which at least takes the following impairing functions: (a) mediate the activity of Rho kinase to phosphorylate MLC and contract actin cytoskeleton. Enhanced contractility of actomyosin subsequently inhibits the function of GJs hemichannel. (b) Affect the sensitivity of MLC to calcium and [Ca2+]i. (c) Downregulate barrier-protective factor Rac1 and cdc42 signaling pathway. The binding of LPS to the Toll-like receptor 4 (TLR4)-MD-2 complex, which depends on the isoform of the Ga subunit (e.g. Gaq/11 or Ga12/13), leads to impaired cAMP signaling following the increase of [Ca2+]i, and causes both the inactivation of Rac1 and the activation of RhoA (for example, via p115RhoGEF). TLR4-mediated activation of receptor on cell surface affects the MLC phosphorylation state through multiple intracellular signaling pathways.

Calcium

Calcium is known as an important second messenger related to hyperpermeability pathways, is essential for maintaining endothelial functions. Many researchers support that GJs is regulated by $[Ca^{2+}]_i$, and addition of Ca^{2+} leads to the close of most of the hemichannel, supposing that GJs takes a role in ALI [50-53]. LPS activates Toll-like receptor 4 (TLR4) and TRPC6-dependent Ca^{2+} signaling to mediate vascular leakage in ALI [54]. Calcium, Rho GTPase and GJs are tightly correlated [35]. The level of $[Ca^{2+}]_i$ is also critical in regulating permeability. Ca^{2+} signals can be propagated along interconnected endothelial cells via GJs, and make inflammation rapidly spread across large areas of the lung tissue [55]. It has been widely proven that increase of $[Ca^{2+}]_i$ affects the activation of RhoA [53]. Disseminated Ca^{2+} through GJs and Rho responses is the mechanistic basis and contributes to the spread of lung microvascular injury in ALI [50-53, 56]. Receptor-mediated Ca^{2+} sensitization through the RhoA kinase pathway regulates the contraction of vascular endothelial cells [57].

HMGB1

High mobility group box 1(HMGB1) protein, a late mediator of sepsis, mediates barrier disruption by cytoskeletal rearrangement, and is essential to paracellular gap formation and barrier dysfunction of ALI [58]. HMGB1-induced lung injury is mediated by multiple surface receptors that including TLR2, TLR4, and the receptor for advanced glycation end products (RAGE) [59, 60]. HMGB1 also contributes directly to the production of proinflammatory cytokines and the development of ALI [9, 60]. Study also proved that in liver I/R injury-induced ALI, HMGB1 increased significantly in both the serum and the lung, while TLR4, p38MAPK and AP-1 signaling pathways were activated simultaneously [61].

Cell Junctions

Endothelial permeability mostly comprises the passage of solutes through paracellular junction, and depends on the integrity of intercellular junctions and actomyosin-based cell contractility. The junctions among the endothelial cells consist of tight junction (TJ), adherens junction (AJ), and gap junction

(GJ) channels. AJs and TJs are spread along the cell border and they function to promote adhesion of opposing cells in the monolayer and to maintain the integrity of the endothelial barrier. TJs are composed of the claudin family, among which claudin 3, 4, and 18 are the most predominantly expressed in the alveolar epithelium [62]. Similarly as those on the epithelium, TJs and AJs on endothelial are also targets of pathological stimuli during ALI [14, 63].

Pulmonary endothelial cells serves as a semi-selective barrier and vital to vessel wall homeostasis and lung function. GJs permit intercellular exchange of molecules at the size about 1.5 kD, and participate in intracellular signaling events that involving the exchange of second messengers (such as ATP, IP_3/Ca^{2+}, cAMP, cGMP) and electrical coupling via current-carrying ionic species. GJs among endothelial cells not only join in the communication of intercellular Ca^{2+} signals, but promote the spatial spread of Ca^{2+} in capillary network to coordinate vascular and endothelial cells contraction [13, 55]. GJs have been found closely related to lung endothelial barrier function [62]. GJs form transmembrane channels between adjacent cells and provide direct electrical and biochemical communications among epithelial cells.

GJs are comprised of a protein family known as connexins (Cxs) [64, 65]. In mammals' lung microvasculature, Cx37, Cx40 and Cx43 are expressed, but Cx40 is a major functional connexin protein presenting in normal PMVECs, and is targeted by LPS-induced signaling in microvascular endothelial cells [66-69]. Expression level of Cx40 intracellular is varying from tissue to cell types, and is the most abundant in microvessels [70]. Cx40 plays a crucial role in the modulation of mouse microcirculation [36, 71]. Modulation of Cx40 expression has been proven plays a role in the development of acute inflammation [66]. LPS treatment down-regulated expression of Cx40 and inhibited function of gap junction by [46, 66, 72, 73] (Fig 1). GJs are regulated by various factors that including trans-junctional voltage, intracellular pH, phosphorylation, and intracellular Ca^{2+} concentration.

Nitric Oxide (NO) and Inducible Nitric Oxide Synthase (iNOS)

NO has been proven to regulate alveolar-capillary barrier [74]. During ALI, it is mainly iNOS that synthesize NO and react to the inducible isoform and sepsis [75, 76]. NO takes variable effects in ALI, such as prompting pulmonary edema, microvascular injury and neutrophilic infiltration in sepsis; however, it sometimes acts as an antimicrobial or anti-inflammatory agent [77]. Furthermore, NO can modulate signaling pathways including NF-κB

[12]. Alveolar macrophages, and alveolar epithelial cells are involved in expressing iNOS in lung in ALI [17]. NO synthesis has been identified in mammals as three distinct genes encoded NOS isozymes: neuronal (nNOS or NOS-1); high-output, calcium-independent, and cytokine-inducible (iNOS or NOS-2); and endothelial (eNOS or NOS-3) calcium–dependent (cNOS) [78]. In the lung, cNOS is constitutively expressed in EC (ecNOS) and in neurones (nNOS). cNOS is important in pulmonary homeostasis, including mediating direct and neurogenic pulmonary vasodilatation, bronchodilation, and immune modulation [74]. In sepsis and ALI, cNOS is proven to be down-regulated [74].

Focal Adhesions (FAs) Complex and FA Kinase (FAK)

Focal adhesions (FAs) complex are assembled at the plasma membrane in discrete regions of integrin-mediated recognition of extracellular matrix (ECM) components. They regulate cytoskeletal organization and play an important role in cell signaling. Endothelial cells are attached to the underlying matrix by FA complexes. Disruption of FA leads to endothelial barrier dysfunction [14, 79]. FA kinase (FAK), the central regulator of FA function, is altered by oxidative stress [10]. FAK spatially and temporally alters activity of RhoA to control cytoskeletal changes and cell motility during cell migration [80]. Silence of FAK has been proven to protect against lung collagen deposition and has been shown the therapeutic potential of FAK [81].

Hyperoxia

During mechanical ventilation, regional alveolar hypoxia contributes to both the pathogenesis and progression of ALI [71]. A recent research has found that a significant increase in the elastic modulus of alveolar epithelial cells following the exposure to hyperoxia, which is correlated with cytoskeletal remodeling through Rho kinase [82, 83]. Rac1 and RhoA play opposing roles in the regulation of hypoxia/reoxygenation-induced permeability in pulmonary artery endothelial cells [38]. Hypoxia leads to damage of alveolar epithelial and endothelial cells [84]. It activates the expression of NF-κB in intestinal epithelial cells, which in turn increases the production of tumor necrosis factor α (TNF-α) but simultaneously attenuates intestinal epithelial apoptosis [84]. Hypoxia causes both the accumulation and retention of pulmonary edema in the alveolar space by impairing junctional complexes and downregulating the Na^+ transporters Na^+-K^+-ATPase and ENaC [14, 85, 86]. Short-term exposure to hyperoxia alveolar type II cells may even induce cell fetal injury [87]. Hyperoxia may also initiate ROS generation and activate inflammasome, leading to altered mechanical properties and apoptosis to alveolar epithelial cells [14, 86, 88].

Prospective Approach to Heal ALI

Low molecular weight heparin inhibits systemic inflammation and prevents endotoxin-induced acute lung injury in rats [89]. Inhibition of RhoA has proven to induce decrease in human umbilical vein endothelial permeability [40]. RhoA has destabilizing effects on endothelial barrier properties, which is contrary to Rac1 or Cdc42. Researchers found that inactivation of NF-κB by andrographolide can protect against LPS-induced ALI [90]. Detection or inhibition of iNOS in mice is protective from LPS-induced mortality [75, 91] through hedgehog signaling pathway [92]. As an important inflammation media, inhibition of extracellular HMGB1 by activation of PPARγ attenuates LPS- and hyperoxia-induced inflammatory in ALI [93, 94].

References

[1] Rubenfeld GD, Caldwell E, Peabody E, et al. Incidence and outcomes of acute lung injury. *N. Engl. J. Med.* 2005; 353:1685-93.
[2] Dixon B, Santamaria JD and Campbell DJ. A phase 1 trial of nebulised heparin in acute lung injury. *Critical Care.* 2008; 12:R64.
[3] Adam O, Lavall D, Theobald K, et al. Rac1-Induced Connective Tissue Growth Factor Regulates Connexin 43 and N-Cadherin Expression in Atrial Fibrillation. *J. Am. Coll Cardiol.* 2010; 55:469-80.
[4] Aird WC. The role of the endothelium in severe sepsis and multiple organ dysfunction syndrome. *Blood. 2003; 101:3765-77.*
[5] Mason RJ. Biology of alveolar type II cells. *Respirology.* 2006; 11 Suppl:S12-5.
[6] Ware LB and Matthay MA. The Acute Respiratory Distress Syndrome. *N. Engl. J. Med.* 2000; 342:1334-49.

[7] Manicone AM. Role of the pulmonary epithelium and inflammatory signals in acute lung injury. *Expet. Rev. Clin. Immunol.* 2009; 5:63-75.
[8] Wang L, Taneja R, Wang W, et al. Human alveolar epithelial cells attenuate pulmonary microvascular endothelial cell permeability under septic conditions. *PLoS One.* 2013; 8:e55311.
[9] Ding N, Wang F, Xiao H, Xu L and She S. Mechanical ventilation enhances HMGB1 expression in an LPS-induced lung injury model. *PLoS One.* 2013; 8:e74633.
[10] Lu Q, Sakhatskyy P, Grinnell K, et al. Cigarette smoke causes lung vascular barrier dysfunction via oxidative stress-mediated inhibition of RhoA and focal adhesion kinase. *Am. J. Physiol. Lung Cell Mol. Physiol.* 2011; 301:L847-57.
[11] Herold S, Mayer K and Lohmeyer J. Acute lung injury: how macrophages orchestrate resolution of inflammation and tissue repair. *Front Immunol.* 2011; 2:65.
[12] D'Alessio FR, Tsushima K, Aggarwal NR, et al. Resolution of experimental lung injury by monocyte-derived inducible nitric oxide synthase. *J. Immunol.* 2012; 189:2234-45.
[13] Bhattacharya J and Matthay MA. Regulation and repair of the alveolar-capillary barrier in acute lung injury. *Annu. Rev. Physiol.* 2013; 75:593-615.
[14] Susanne Herold, Nieves M. Gabrielli and Vadász I. Novel concepts of acute lung injury and alveolar-capillary barrier dysfunction. *Am. J. Physiol. Lung Cell Mol. Physiol.* 2013; 305:L665–L81.
[15] Sznajder JI. Strategies to increase alveolar epithelial fluid removal in the injured lung. *Am. J. Respir. Crit. Care Med.* 1999; 160:1441-2.
[16] Alexander Zarbock and Ley K. The role of platelets in acute lung injury (ALI). *Front Biosci.* 2009; 14:150-8.
[17] Johnson ER and MA M. Acute Lung Injury: Epidemiology, Pathogenesis, and Treatment. *J. Aerosol. Med. Pulm. Drug Deliv.* 2010; 23:243-52.
[18] Bastacky J, Lee CY, Goerke J, et al. Alveolar lining layer is thin and continuous: low-temperature scanning electron microscopy of rat lung. *J. Appl. Physiol.* 1995; 79:1615-26.
[19] Vadasz I, Weiss CH and Sznajder JI. Ubiquitination and proteolysis in acute lung injury. *Chest.* 2012; 141:763-71.
[20] Pugin J, Verghese G, Widmer MC and Matthay MA. The alveolar space is the site of intense inflammatory and profibrotic reactions in the early

phase of acute respiratory distress syndrome. *Crit. Care Med.* 1999; 27:304-12.
[21] Grommes J and Soehnlein O. Contribution of neutrophils to acute lung injury. *Mol. Med.* 2011; 17:293-307.
[22] Narasaraju T, Yang E, Samy RP, et al. Excessive neutrophils and neutrophil extracellular traps contribute to acute lung injury of influenza pneumonitis. *Am. J. Pathol.* 2011; 179:199-210.
[23] Saffarzadeh M, Juenemann C, Queisser MA, et al. Neutrophil extracellular traps directly induce epithelial and endothelial cell death: a predominant role of histones. *PLoS One.* 2012; 7:e32366.
[24] Matthay MA. Acute Lung Injury and the Acute Respiratory Distress Syndrome: Four Decades of Inquiry into Pathogenesis and Rational Management. *Am. J. Resp. Cell Mol.* 2005; 33:319-27.
[25] Cavnar PJ, Berthier E, Beebe DJ and Huttenlocher A. Hax1 regulates neutrophil adhesion and motility through RhoA. *J. Cell Biol.* 2011; 193:465-73.
[26] Beaulieu LM and Freedman JE. The role of inflammation in regulating platelet production and function: Toll-like receptors in platelets and megakaryocytes. *Thromb Res.* 2010; 125:205-9.
[27] Janssens S and Beyaert R. Role of Toll-Like Receptors in Pathogen Recognition. *Clin. Microbiol. Rev.* 2003; 16:637-46.
[28] Kuper C, Beck FX and Neuhofer W. Toll-like receptor 4 activates NF-kappaB and MAP kinase pathways to regulate expression of proinflammatory COX-2 in renal medullary collecting duct cells. *Am. J. Physiol. Renal Physiol.* 2012; 302:F38-46.
[29] Chen Z, Zhang X, Chu X, et al. Preventive effects of valnemulin on lipopolysaccharide-induced acute lung injury in mice. *Inflammation.* 2010; 33:306-14.
[30] Michael Karin and Ben-Neriah Y. Phosphorylation Meets Ubiquitination: The Control of NF-κB Activity. *Ann. Rev. Immunol.* 2000; 18:621-33.
[31] Vohwinkel CU, Lecuona E, Sun H, et al. Elevated CO(2) levels cause mitochondrial dysfunction and impair cell proliferation. *J. Biol. Chem.* 2011; 286:37067-76.
[32] Tan W, Palmby TR, Gavard J, Amornphimoltham P, Zheng Y and Gutkind JS. An essential role for Rac1 in endothelial cell function and vascular development. *The FASEB Journal.* 2008; 22:1829-38.
[33] Wojciak-Stothard B and Ridley AJ. Rho GTPases and the regulation of endothelial permeability. *Vasc. Pharmacol.* 2002; 39:187-99.

[34] Beata Wojciak-Stothard, Lillian Yen Fen Tsang, Ewa Paleolog, Susan M. Hall and Sheila and Haworth G. Rac1 and RhoA as regulators of endothelial phenotype and barrier function in hypoxia-induced neonatal pulmonary hypertension. *Am. J. Physiol. Lung Cell Mol. Physiol.* 2006; 290:L1173-L82.

[35] Derangeon M, Bourmeyster N, Plaisance I, et al. RhoA GTPase and F-actin dynamically regulate the permeability of Cx43-made channels in rat cardiac myocytes. *J. Biol. Chem.* 2008; 283:30754-65.

[36] Haefliger JA, Nicod P and Meda P. Contribution of connexins to the function of the vascular wall. *Cardiovasc. Res.* 2004; 62:345-56.

[37] Citi S, Spadaro D, Schneider Y, Stutz J and Pulimeno P. Regulation of small GTPases at epithelial cell-cell junctions. *Mol. Membr. Biol..* 2011; 28:427-44.

[38] Wojciak-Stothard B, Tsang LY and Haworth SG. Rac and Rho play opposing roles in the regulation of hypoxia/reoxygenation-induced permeability changes in pulmonary artery endothelial cells. *Am. J. Physiol. Lung Cell Mol. Physiol.* 2005; 288:L749-60.

[39] Ramchandran R, Mehta D, Vogel SM, Mirza MK, Kouklis P and Malik AB. Critical role of Cdc42 in mediating endothelial barrier protection in vivo. *Am. J. Physiol. Lung Cell Mol. Physiol.* 2008; 295:L363-9.

[40] Waschke J, Burger S, Curry FR, Drenckhahn D and Adamson RH. Activation of Rac-1 and Cdc42 stabilizes the microvascular endothelial barrier. *Histochem. Cell Biol.* 2006; 125:397-406.

[41] Van Aelst L and D'Souza-Schorey C. Rho GTPases and signaling networks. *Genes & Development.* 1997; 11:2295-322.

[42] Wettschureck N and Offermanns S. Rho/Rho-kinase mediated signaling in physiology and pathophysiology. *J. Mol. Med.* 2002; 80:629-38.

[43] Wojciak-Stothard B Fau - Tsang LYF, Tsang Ly Fau - Paleolog E, Paleolog E Fau - Hall SM, Hall Sm Fau - Haworth SG and Haworth SG. Rac1 and RhoA as regulators of endothelial phenotype and barrier function in hypoxia-induced neonatal pulmonary hypertension. *Am. J. Physiol. Lung Cell Mol. Physiol.* 2006; 290:L1173-82.

[44] Beckers CML, van Hinsbergh VWM and van Nieuw Amerongen GP. Driving Rho GTPase activity in endothelial cells regulates barrier integrity. *Thromb Haemost..* 2009; 103:40-55.

[45] Etienne-Manneville S HA. Rho GTPases in cell biology. *Nature.* 2002; 420:629-35.

[46] Tohru Minamino IK. Regeneration of the endothelium as a novel therapeutic strategy for acute lung injury. *J. Clin. Invest.* 2006; 116:2316-9.
[47] Spindler V, Schlegel N and Waschke J. Role of GTPases in control of microvascular permeability. *Cardiovasc. Res.* 2010; 87:243-53.
[48] Wettschureck N and Offermanns S. Rho/Rho-kinase mediated signaling in physiology and pathophysiology. *J. Mol. Med.* (Berl). 2002; 80:629-38.
[49] Raf Ponsaerts CDh, Fréderic Hertens, Jan B. Parys, Luc Leybaert, Johan Vereecke, Bernard Himpens, Geert Bultynck. RhoA GTPase Switch Controls Cx43-Hemichannel Activity through the Contractile System. *Plos One.* 2012; 7:e420774.
[50] Angus DC. Epidemiology of severe sepsis in the United States: analysis of incidence, outcome, and associated costs of care. *Crit. Care Med.* 2001; 29:1303-10.
[51] Alberti C. Epidemiology of sepsis and infection in ICU patients from an international multicentre cohort study. *Intensive Care Med.* 2002; 28:108-21.
[52] Poltorak A. Defective LPS signaling in C3H/HeJ and C57BL/10ScCr mice: mutations in Tlr4 gene. *Science.* 1998; 282:2085-8.
[53] Masiero L Fau - Lapidos KA, Lapidos Ka Fau - Ambudkar I, Ambudkar I Fau - Kohn EC and Kohn EC. Regulation of the RhoA pathway in human endothelial cell spreading on type IV collagen: role of calcium influx. *J. Cell Sci.* 1999 Oct; 112:3205-13.
[54] Tauseef M, Knezevic N, Chava KR, et al. TLR4 activation of TRPC6-dependent calcium signaling mediates endotoxin-induced lung vascular permeability and inflammation. *J. Exp. Med.* 2012; 209:1953-68.
[55] Parthasarathi K, Ichimura H, Monma E, et al. Connexin 43 mediates spread of Ca2+-dependent proinflammatory responses in lung capillaries. *J. Clin. Invest.* 2006; 116:2193-200.
[56] Zhu L, Qi XY, Aoudjit L, et al. Nuclear factor of activated T-cells mediates RhoA-induced fibronectin upregulation in glomerular podocytes. *Am. J. Physiol. Renal. Physiol.* 2013:Epub ahead of print.
[57] Sims SM, Chrones T and Preiksaitis HG. Calcium sensitization in human esophageal muscle: role for RhoA kinase in maintenance of lower esophageal sphincter tone. *J. Pharmacol. Exp. Ther.* 2008; 327:178-86.

[58] Rachel K. Wolfson, Eddie T. Chiang and Garcia JGN. HMGB1 induces human lung endothelial cell cytoskeletal rearrangement and barrier disruption. *Microvas. Res.* 2011; 81:189-97.

[59] Sims GP, Rowe DC, Rietdijk ST, Herbst R and Coyle AJ. HMGB1 and RAGE in inflammation and cancer. *Ann. Rev. Immunol.* 2010; 28:367-88.

[60] Deng Y, Yang Z, Gao Y, et al. Toll-like receptor 4 mediates acute lung injury induced by high mobility group box-1. *PLoS One.* 2013; 8:e64375.

[61] Yang Z, Deng Y, Su D, et al. TLR4 as receptor for HMGB1-mediated acute lung injury after liver ischemia/reperfusion injury. *Lab. Invest.* 2013; 93:792-800.

[62] Koval M. Claudin heterogeneity and control of lung tight junctions. *Ann. Rev. Physiol.* 2013; 75:551-67.

[63] Vandenbroucke E, Mehta D, Minshall R and Malik AB. Regulation of endothelial junctional permeability. *Ann. N Y Acad. Sci.* 2008; 1123:134-45.

[64] Figueroa XF and Duling BR. Gap Junctions in the Control of Vascular Function. *Antioxid. Redox Sign .* 2009; 11:251-66.

[65] Evans WH MP. Gap junctions: structure and function (Review). *Mol Membr. Biol.* 2002; 19:121-36.

[66] Rignault S Fau - Haefliger J-A, Haefliger Ja Fau - Waeber B, Waeber B Fau - Liaudet L, Liaudet L Fau - Feihl F and Feihl F. Acute inflammation decreases the expression of connexin 40 in mouse lung. *Shock.* 2007; 28:78-85.

[67] Bastide B, Neyses L, Ganten D, Paul M, Willecke K and Traub O. Gap junction protein connexin40 is preferentially expressed in vascular endothelium and conductive bundles of rat myocardium and is increased under hypertensive conditions. *Circ. Res.* 1993; 73:1138-49.

[68] Bolon ML, Kidder GM, Simon AM and Tyml K. Lipopolysaccharide reduces electrical coupling in microvascular endothelial cells by targeting connexin40 in a tyrosine-, ERK1/2-, PKA-, and PKC-dependent manner. *J. Cell Physiol.* 2007; 211:159-66.

[69] Rignault S, Haefliger JA, Gasser D, et al. Sepsis up-regulates the expression of connexin 40 in rat aortic endothelium. *Crit. Care Med.* 2005; 33:1302-10.

[70] Filep JG. Sepsis and vascular dysfunction Connections with endothelial Cx40. *Crit. Care Med.* 2005; 33:1442-3.

[71] de Wit C, Roos F, Bolz SS, et al. Impaired Conduction of Vasodilation Along Arterioles in Connexin40-Deficient Mice. *Circ. Res.* 2000; 86:649-55.
[72] Zhang J, Wang W, Sun J, et al. Gap Junction Channel Modulates Pulmonary Vascular Permeability through Calcium in Acute Lung Injury: An Experimental Study. *Respiration.* 2010; 80:236-45.
[73] Simon AM, McWhorter AR, Chen H, Jackson CL and Ouellette Y. Decreased intercellular communication and connexin expression in mouse aortic endothelium during lipopolysaccharide-induced inflammation. *J. Vasc. Res.* 2004; 41:323-33.
[74] Mehta S. The effects of nitric oxide in acute lung injury. *Vascul. Pharmacol.* 2005; 43:390-403.
[75] Razavi HM, Werhun R, Scott JA, et al. Effects of inhaled nitric oxide in a mouse model of sepsis-induced acute lung injury. *Crit. Care Med.* 2002; 30:868-73.
[76] Ermert M, Ruppert C, Gunther A, Duncker HR, Seeger W and Ermert L. Cell-specific nitric oxide synthase-isoenzyme expression and regulation in response to endotoxin in intact rat lungs. *Lab Invest. 2002; 82:425-41.*
[77] Bogdan C, Rollinghoff M and Diefenbach A. The role of nitric oxide in innate immunity. *Immunol. Rev.* 2000; 173:17-26.
[78] Knowles RG and Moncada S. Nitric oxide synthases in mammals. *Biochem. J.* 1994; 298 (Pt 2):249-58.
[79] Zebda N, Dubrovskyi O and Birukov KG. Focal adhesion kinase regulation of mechanotransduction and its impact on endothelial cell functions. *Microvasc. Res.* 2012; 83:71-81.
[80] Schaller MD. Cellular functions of FAK kinases: insight into molecular mechanisms and novel functions. *J. Cell Sci.* 2010; 123:1007-13.
[81] Petroni RC, Teodoro WR, Guido MC, et al. Role of focal adhesion kinase in lung remodeling of endotoxemic rats. *Shock.* 2012; 37:524-30.
[82] Roan E, Wilhelm K, Bada A, et al. Hyperoxia alters the mechanical properties of alveolar epithelial cells. *Am. J. Physiol. Lung Cell Mol. Physiol.* 2012; 302:L1235-41.
[83] Wilhelm KR, Roan E, Ghosh MC, Parthasarathi K and Waters CM. Hyperoxia increases the elastic modulus of alveolar epithelial cells through Rho kinase. *FEBS J.* 2014; 281:957-69.
[84] Holger K. Eltzschig and Carmeliet P. Hypoxia and inflammation. *N. Engl. J. Med.* 2011; 364:656-65.
[85] Caraballo JC, Yshii C, Butti ML, et al. Hypoxia increases transepithelial electrical conductance and reduces occludin at the plasma membrane in

alveolar epithelial cells via PKC-zeta and PP2A pathway. *Am. J. Physiol. Lung Cell Mol. Physiol.* 2011; 300:L569-78.

[86] Dada LA, Chandel NS, Ridge KM, Pedemonte C, Bertorello AM and Sznajder JI. Hypoxia-induced endocytosis of Na,K-ATPase in alveolar epithelial cells is mediated by mitochondrial reactive oxygen species and PKC-ζ. *J. Clin. Invest.* 2003; 111:1057-64.

[87] Lee H-S. Fetal Alveolar Type II Cell Injury Induced by Short-term Exposure to Hyperoxia. *Neonatal Medicine.* 2013; 20:300.

[88] Kallet RH and Matthay MA. Hyperoxic Acute Lung Injury. *Respiratory Care.* 2012; 58:123-41.

[89] Luan Z-G, Naranpurev M and Ma X-C. Treatment of Low Molecular Weight Heparin Inhibits Systemic Inflammation and Prevents Endotoxin-Induced Acute Lung Injury in Rats. *Inflammation.* 2014:1-9.

[90] Zhu T, Wang DX, Zhang W, et al. Andrographolide protects against LPS-induced acute lung injury by inactivation of NF-kappaB. *PLoS One.* 2013; 8:e56407.

[91] Gao P, Yang X, Mungur L, Kampo S and Wen Q. Adipose tissue-derived stem cells attenuate acute lung injury through eNOS and eNOS-derived NO. *Int. J. Mol. Med.* 2013; 31:1313-8.

[92] Yang Y, Li Q, Deng Z, et al. Protection from lipopolysaccharide-induced pulmonary microvascular endothelial cell injury by activation of hedgehog signaling pathway. *Mol. Biol. Rep.* 2011; 38:3615-22.

[93] Wang G, Liu L, Zhang Y, et al. Activation of PPARgamma attenuates LPS-induced acute lung injury by inhibition of HMGB1-RAGE levels. *Eur. J. Pharmacol.* 2014; 726C:27-32.

[94] Entezari M, Javdan M, Antoine DJ, et al. Inhibition of extracellular HMGB1 attenuates hyperoxia-induced inflammatory acute lung injury. *Redox Biol.* 2014; 2:314-22.

In: Acute Lung Injury
Editor: Daniela Mokrá
ISBN: 978-1-61470-426-3
© 2014 Nova Science Publishers, Inc.

Chapter 3

PHARMACOLOGICAL THERAPIES FOR ACUTE RESPIRATORY DISTRESS SYNDROME

Hironobu Hamada[*]

Department of Physical Analysis and Therapeutic Sciences,
Graduate School of Biomedical and Health Sciences,
Hiroshima University, Japan

ABSTRACT

Although many pharmacological therapies, including corticosteroids, neutrophil elastase inhibitors, and anticoagulants, have been evaluated for use in patients with acute respiratory distress syndrome (ARDS), most of these have failed to reduce mortality or improve patient outcomes. However, some therapies have shown beneficial effects for certain subgroups of patients. This review provides a summary of pharmacological therapies used for patients with ARDS.

Keywords: Acute respiratory distress syndrome, corticosteroids, neutrophil elastase inhibitors, anticoagulants, pharmacological therapy

---

[*] Corresponding Author address: Hironobu Hamada, M.D., Ph.D., Department of Physical Analysis and Therapeutic Sciences, Graduate School of Biomedical and Health Sciences, Hiroshima University, 1-2-3 Kasumi, Minami-ku, Hiroshima 734-8551, Japan Email: hirohamada@hiroshima-u.ac.jp.

INTRODUCTION

Acute respiratory distress syndrome (ARDS) is characterized by acute onset, severe hypoxia, and bilateral pulmonary infiltration with no increase in the pulmonary artery wedge pressure [1]. Acute lung injury (ALI) is defined as a PaO_2/F_IO_2 ratio of < 300 mmHg and ARDS is defined as a more severe condition with a PaO_2/F_IO_2 ratio of < 200 mmHg. The pathogenesis of ARDS includes pulmonary injury and edema that result from the excess production of inflammatory mediators [2].

ARDS was first reported in 1967 [3] and was defined as adult respiratory distress syndrome in 1971 [4]. Although numerous clinical trials were conducted during the last 40 years, no effective pharmacological therapies have been reported that improved the survival of patients with ARDS. This review focuses on the pharmacological therapies currently used for ARDS.

CORTICOSTEROIDS

Corticosteroids have been considered an ideal treatment for ARDS because of their potent anti-inflammatory effects. Corticosteroids inhibit the production of pro-inflammatory cytokines and chemokines and associated adhesion molecules. In addition, corticosteroids inhibit fibroblast proliferation, reduce collagen deposition and neutrophil activation, and also induce inflammatory cell apoptosis [5].

Several corticosteroid regimens have been investigated for patients with ARDS. High-dose methylprednisolone (30 mg/kg every six hours for 24 hours) did not alter the outcomes of patients with early-phase ARDS [6]. There were no significant differences between the methylprednisolone and placebo groups in terms of mortality or in the reversal of ARDS by 45 days. Infectious complication rates were also similar in the methylprednisolone and placebo groups.

Administering high-dose methylprednisolone also did not prevent the development of ARDS and increased the infection rates and mortality rates of high risk patients [7-9]. High-dose methylprednisolone (30 mg/kg) was administered every six hours for 24 hours to patients at risk of ARDS due to sepsis [7]. The incidence of ARDS increased in the methylprednisolone group as compared to the placebo group (32% vs. 25%), although this difference was not statistically significant. Significantly fewer methylprednisolone treated patients had ARDS reversal as compared to those treated with a placebo. The 14-day mortality rate of patients with ARDS treated with methylprednisolone was significantly higher than that of those treated with a placebo. High-dose methylprednisolone (30 mg/kg) was also administrated every six hours for 48 hours to mechanically ventilated patients at risk of ARDS [9]. The incidence of ARDS and early infectious complications were significantly higher in the methylprednisolone group as compared to the placebo group.

Low-dose corticosteroids might improve the outcomes of patients with early-phase ALI/ARDS [10-12]. Low-dose methylprednisolone (1 mg/kg) was first administrated to patients with ARDS for 14 days, followed by 0.5 mg/kg for 7 days and 0.25 mg/kg for 7 days [11]. The methylprednisolone treated group had twice as many patients with a 1-point reduction in lung injury scores (LIS) and breathing without assistance as compared to those treated with a placebo. Methylprednisolone treated patients had significantly lower LIS and multiple organ dysfunction syndrome (MODS) scores by day 7. Low-dose methylprednisolone treatment was significantly associated with a reduction in the duration of mechanical ventilation, intensive care unit (ICU) stays, and ICU mortality. Treated patients also had a lower rate of infections.

A meta-analysis of five cohort studies and four randomized controlled trials was performed to determine whether low-dose corticosteroids (e.g., 0.5-2.5 mg/kg of methylprednisolone or equivalent) could reduce the mortality and morbidity associated with ALI/ARDS without increasing the risk of adverse events [12]. The duration of treatment was from 7 to 32 days (mean of 8 days). Using low-dose corticosteroids was significantly associated with a reduced risk of mortality and an improvement in all morbidity outcomes, including the duration of mechanical ventilation, MODS scores, and PaO_2/F_1O_2 ratios. There were no increases in any major complications, including infections and neuromyopathy, among low-dose corticosteroid treated patients. However, more studies will be needed to determine the effects of low-dose corticosteroids for a larger number of ALI/ARDS patients, as the studies cited only enrolled small numbers of patients.

NEUTROPHIL ELASTASE INHIBITORS

Neutrophils accumulate in the lungs of patients with ARDS. Neutrophil elastase is produced in the lungs [13] and plays a significant role in the pathogenesis of ARDS, including pulmonary vascular hyperpermeability [14]. Intratracheal administration of neutrophil elastase caused lung injuries in mice

[15]. By comparison, no lung injuries were found in animals whose neutrophil levels were reduced [16]. In addition, a neutrophil elastase inhibitor, sivelestat, reduced the numbers of deaths among hamsters associated with severe lung injury after hydrochloric acid aspiration [17].

Two phase III and IV studies conducted in Japan have shown the clinical usefulness of sivelestat for ALI/ARDS patients [18, 19]. In a double-blind phase III study, administering sivelestat significantly increased pulmonary function improvement ratings, reduced the duration of mechanical ventilation, and shortened ICU stays, although there was no significant benefit with regard to the survival rate. In a phase IV non-randomized study, the adjusted mean number of ventilator free days (VFD), the ventilator-weaning rate, and the ICU discharge rate were significantly higher for a sivelestat group as compared to the control group. The adjusted 180-day survival rate was significantly higher for the sivelestat group than for the control group. However, the Sivelestat Trial in ALI Patients Requiring Mechanical Ventilation (STRIVE) study failed to demonstrate any efficacy for sivelestat [20], as sivelestat use did not affect the primary end points of VFD and 28-day all-cause mortality. There were no differences in adverse events or serious adverse events between treatment and control groups.

It was recently reported that sivelestat was effective for patients with ALI/ARDS and sepsis in several retrospective studies [21-23]. One study assessed septic patients associated with ARDS and disseminated intravascular coagulation (DIC) [21]. The mean ICU stay for the sivelestat group was significantly shorter than that for the control group. Sivelestat administration was an independent predictor of survival for septic patients associated with both ARDS and DIC.

The efficacy of sivelestat for patients with ALI/ARDS associated with abdominal sepsis was also examined [22]. The sivelestat group had significant improvements in oxygenation, thrombocytopenia, and MODS scores. The number of ventilator days and ICU stays were also significantly lower in the sivelestat group, and the hospital mortality rate decreased by half in the sivelestat group.

In addition, the clinical features of patients with ALI/ARDS that might affect the efficacy of sivelestat were investigated [23]. Sivelestat was more effective for ALI/ARDS patients with PaO_2/F_IO_2 ratios of $\geq$ 140 mmHg or sepsis. Sivelestat use significantly prolonged the survival of and resulted in higher VFD and increased changes in PaO_2/F_IO_2 ratios among septic ALI/ARDS patients. Larger prospective studies will be needed to determine the effects of sivelestat for ALI/ARDS patients with sepsis.

ANTICOAGULANT THERAPIES

ALI/ARDS is associated with increased procoagulant and reduced fibrinolytic activities in alveolar and interstitial spaces in the lung, which is the rationale for using anticoagulants for treating ALI/ARDS. Several animal studies demonstrated that administering anticoagulants, including tissue factor pathway inhibitor (TFPI), anti-thrombin, activated protein C (APC), and soluble thrombomodurin, improved lung injury due to their anti-inflammatory properties, although clinical studies have not shown any positive results [24].

Treatment with recombinant TFPI had no effect on all-cause mortality for patients with severe sepsis and high international normalized ratios [25]. In addition, high-dose anti-thrombin III therapy administrated within 6 hours after onset (30 000 IU total over 4 days) did not provide any survival advantage for adult patients with severe sepsis and septic shock [26]. There were no significant differences in 28- and 90-day mortality rates, lengths of ICU stays, and the occurrence of new organ dysfunction between the anti-thrombin III treatment and placebo groups.

APC is an endogenous protein that promotes fibrinolysis and inhibits thrombosis and inflammation and is an important modulator of coagulation and inflammation associated with severe sepsis [27]. Recombinant human APC significantly reduced the mortality of patients with severe sepsis [28]. However, a recent study reported that recombinant human APC did not significantly reduce the mortality at 28 or 90 days for patients with septic shock [29]. Furthermore, recombinant human APC did not improve the outcomes for ALI/ARDS patients [30], as there were no statistically significant differences in the numbers of ventilator-free days or mortality at 60 days between the placebo and APC groups.

CONCLUSION

There has been little success in developing effective pharmacological therapies to reduce mortality or improve the outcomes of patients with ARDS, even though numerous pharmacological therapies have been evaluated. Additional clinical studies will be needed to develop agents that are useful for patients with ARDS or to identify certain subgroups of ARDS patients for which some therapies have shown beneficial effects.

REFERENCES

[1] Bernard GR, Artigas A, Brigham KL, Carlet J, Falke K, Hudson L, Lamy M, Legall JR, Morris A, Spragg R. The American-European Consensus Conference on ARDS. Definitions, mechanisms, relevant outcomes, and clinical trial coordination. *Am. J. Respir. Crit. Care Med.* 1994; 149: 818-24.

[2] Ware LB, Matthay MA. The acute respiratory distress syndrome. *N. Engl. J. Med.* 2004; 342: 1334-49.

[3] Ashbaugh DG, Bigelow DB, Petty TL, Levine BE. Acute respiratory distress in adults. *Lancet* 1967; 2: 319-23.

[4] Petty TL, Ashbaugh DG. The adult respiratory distress syndrome. Clinical features, factors influencing prognosis and principles of management. *Chest* 1971; 60: 233-9.

[5] Thompson BT. Glucocorticoids and acute lung injury. *Crit. Care Med.* 2003; 31: S253-7.

[6] Bernard GR, Luce JM, Sprung CL, Rinaldo JE, Tate RM, Sibbald WJ, Kariman K, Higgins S, Bradley R, Metz CA, Harris TR, Brigham KL. High-dose corticosteroids in patients with the adult respiratory distress syndrome. *N. Engl. J. Med.* 1987; 317: 1565-70.

[7] Bone RC, Fisher CJ Jr, Clemmer TP, Slotman GJ, Metz CA; Methylprednisolone Severe Sepsis Study Group. Early methylprednisolone treatment for septic syndrome and the adult respiratory distress syndrome. *Chest* 1987; 92: 1032-6.

[8] Luce JM, Montgomery AB, Marks JD, Turner J, Metz CA, Murray JF. Ineffectiveness of high-dose methylprednisolone in preventing parenchymal lung injury and improving mortality in patients with septic shock. *Am. Rev. Respir. Dis.* 1988; 138: 62-8.

[9] Weigelt JA, Norcross JF, Borman KR, Snyder WH 3rd. Early steroid therapy for respiratory failure. *Arch. Surg.* 1985; 120: 536-40.

[10] Annane D, Sébille V, Bellissant E; Ger-Inf-05 Study Group. Effect of low doses of corticosteroids in septic shock patients with or without early acute respiratory distress syndrome. *Crit. Care Med.* 2006; 34: 22-30.

[11] Meduri GU, Golden E, Freire AX, Taylor E, Zaman M, Carson SJ, Gibson M, Umberger R. Methylprednisolone infusion in early severe ARDS: results of a randomized controlled trial. *Chest* 2007; 131: 954-63.

[12] Tang BM, Craig JC, Eslick GD, Seppelt I, McLean AS. Use of corticosteroids in acute lung injury and acute respiratory distress syndrome: a systematic review and meta-analysis. *Crit. Care Med.* 2009; 37: 1594-603.
[13] McGuire WW, Spragg RG, Cohen AB, Cochrane CG. Studies on the pathogenesis of the adult respiratory distress syndrome. *J. Clin. Invest.* 1982; 69: 543-53.
[14] Repine JE. Scientific perspectives on adult respiratory distress syndrome. *Lancet* 1992; 339: 466-9.
[15] Janoff A, White R, Carp H, Harel S, Dearing R, Lee D. Lung injury induced by leukocytic proteases. *Am. J. Pathol.* 1979; 97: 111-36.
[16] Ghio AJ, Kennedy TP, Hatch GE, Tepper JS. Reduction of neutrophil influx diminishes lung injury and mortality following phosgene inhalation. *J. Appl. Physiol.* 1991; 71: 657-65.
[17] Hagio T, Matsumoto S, Nakao S, Abiru T, Ohno H, Kawabata K. Elastase inhibition reduced death associated with acid aspiration-induced lung injury in hamsters. *Eur. J. Pharmacol.* 2004; 488: 173-80.
[18] Tamakuma S, Ogawa M, Aikawa N, Kubota T, Hirasawa H, Ishizaka A, Taenaka N, Hamada C, Matsuoka S, Abiru T. Relationship between neutrophil elastase and acute lung injury in humans. *Pulm. Pharmacol. Ther* 2004; 17: 271-9.
[19] Aikawa N, Ishizaka A, Hirasawa H, Shimazaki S, Yamamoto Y, Sugimoto H, Shinozaki M, Taenaka N, Endo S, Ikeda T, Kawasaki Y. Reevaluation of the efficacy and safety of the neutrophil elastase inhibitor, Sivelestat, for the treatment of acute lung injury associated with systemic inflammatory response syndrome; a phase IV study. *Pulm. Pharmacol. Ther.* 2011; 24: 549-54.
[20] Zeiher BG, Artigas A, Vincent JL, Dmitrienko A, Jackson K, Thompson BT, Bernard G; STRIVE Study Group. Neutrophil elastase inhibition in acute lung injury: results of the STRIVE study. *Crit. Care Med.* 2004; 32: 1695-702.
[21] Hayakawa M, Katabami K, Wada T, Sugano M, Hoshino H, Sawamura A, Gando S. Sivelestat (selective neutrophil elastase inhibitor) improves the mortality rate of sepsis associated with both acute respiratory distress syndrome and disseminated intravascular coagulation patients. *Shock* 2010; 33: 14-8.
[22] Tsuboko Y, Takeda S, Mii S, Nakazato K, Tanaka K, Uchida E, Sakamoto A. Clinical evaluation of sivelestat for acute lung injury/acute

respiratory distress syndrome following surgery for abdominal sepsis. *Drug Des. Devel. Ther.* 2012; 6: 273-8.

[23] Miyoshi S, Hamada H, Ito R, Katayama H, Irifune K, Suwaki T, Nakanishi N, Kanematsu T, Dote K, Aibiki M, Okura T, Higaki J. Usefulness of a selective neutrophil elastase inhibitor, sivelestat, in acute lung injury patients with sepsis. *Drug Des. Devel. Ther.* 2013; 7: 305-15.

[24] Laterre PF, Wittebole X, Dhainaut JF. Anticoagulant therapy in acute lung injury. *Crit. Care Med.* 2003; 31: S329-36.

[25] Abraham E, Reinhart K, Opal S, Demeyer I, Doig C, Rodriguez AL, Beale R, Svoboda P, Laterre PF, Simon S, Light B, Spapen H, Stone J, Seibert A, Peckelsen C, De Deyne C, Postier R, Pettilä V, Artigas A, Percell SR, Shu V, Zwingelstein C, Tobias J, Poole L, Stolzenbach JC, Creasey AA; OPTIMIST Trial Study Group. Efficacy and safety of tifacogin (recombinant tissue factor pathway inhibitor) in severe sepsis: a randomized controlled trial. *JAMA* 2003; 290: 238-47.

[26] Warren BL, Eid A, Singer P, Pillay SS, Carl P, Novak I, Chalupa P, Atherstone A, Pénzes I, Kübler A, Knaub S, Keinecke HO, Heinrichs H, Schindel F, Juers M, Bone RC, Opal SM; KyberSept Trial Study Group. High-dose antithrombin III in severe sepsis: a randomized controlled trial. *JAMA* 2001; 286: 1869-78.

[27] Esmon CT. The protein C anticoagulant pathway. *Arterioscler. Thromb.* 1992; 12: 135-45.

[28] Bernard GR, Vincent JL, Laterre PF, LaRosa SP, Dhainaut JF, Lopez-Rodriguez A, Steingrub JS, Garber GE, Helterbrand JD, Ely EW, Fisher CJ Jr; Recombinant human protein C Worldwide Evaluation in Severe Sepsis (PROWESS) study group. Efficacy and safety of recombinant human activated protein C for severe sepsis. *N. Engl. J. Med.* 2001; 344: 699-709.

[29] Ranieri VM, Thompson BT, Barie PS, Dhainaut JF, Douglas IS, Finfer S, Gårdlund B, Marshall JC, Rhodes A, Artigas A, Payen D, Tenhunen J, Al-Khalidi HR, Thompson V, Janes J, Macias WL, Vangerow B, Williams MD; PROWESS-SHOCK Study Group. Drotrecogin alfa (activated) in adults with septic shock. *N. Engl. J. Med.* 2012; 366: 2055-64.

[30] Liu KD, Levitt J, Zhuo H, Kallet RH, Brady S, Steingrub J, Tidswell M, Siegel MD, Soto G, Peterson MW, Chesnutt MS, Phillips C, Weinacker A, Thompson BT, Eisner MD, Matthay MA. Randomized clinical trial of activated protein C for the treatment of acute lung injury. *Am. J. Respir. Crit. Care Med.* 2008; 178: 618-23.

In: Acute Lung Injury
Editor: Daniela Mokrá

ISBN: 978-1-61470-426-3
© 2014 Nova Science Publishers, Inc.

Chapter 4

NOVEL TRENDS IN PHARMACOTHERAPY OF ACUTE LUNG INJURY AND ACUTE RESPIRATORY DISTRESS SYNDROME

Daniela Mokrá[1] and Juraj Mokry[2]*

[1]Department of Physiology, [2]Department of Pharmacology, Comenius University in Bratislava, Jessenius Faculty of Medicine in Martin, Martin, Slovakia, EU

ABSTRACT

Acute respiratory distress syndrome (ARDS) and its milder form acute lung injury (ALI) occur as a result of various diseases and situations including sepsis, pneumonia, trauma, acute pancreatitis, aspiration of gastric contents, near-drowning etc. ALI/ARDS is characterized by diffuse alveolar damage, alveolar capillary leakage, lung edema, neutrophil-derived inflammation, and surfactant dysfunction. These changes lead to clinical manifestations of decreased lung compliance, severe hypoxemia, and bilateral pulmonary infiltrates. Appropriate treatment involves protective lung ventilation, optimal fluid management, and pharmacological treatment. Despite there are several possibilities of potentially beneficial pharmacotherapy of ALI/ARDS, such as pulmonary

---

* Corresponding Author address: Daniela Mokrá, MD., PhD., Department of Physiology, Comenius University in Bratislava, Jessenius Faculty of Medicine in Martin, Mala Hora 4, Martin, SK-03601 Martin, Slovakia, EU. Email: mokra@jfmed.uniba.sk; Supported by: APVV-0435-11, VEGA 1/0305/14, VEGA 1/0260/14, BioMed (ITMS 26220220187).

vasodilators, corticosteroids, antioxidants, methylxanthines, or exogenous surfactant, the results of experimental and clinical studies are often controversial. This chapter reviews the use of various pharmacological agents and critically evaluates their effects in animal models and in the patients with ALI/ARDS.

Keywords: Acute respiratory distress syndrome, acute lung injury, corticoisteroids, statins, surfactant, vasodilators, prostacyclin, antioxidants, methylxanthines, gene-therapy, cell therapy, nitric oxide, myorelaxants, beta-agonists

DEFINITION OF ALI/ARDS

Acute respiratory distress syndrome (ARDS) is a frequent cause of morbidity and mortality in critically ill patients, with mortality as high as 50% [1].

One of older definitions of ARDS was defined by the American-European Consensus Conference (AECC) [2], where it is characterized by:

- acute onset
- bilateral infiltrates on chest radiography
- pulmonary artery wedge pressure ≤18 mm Hg or the absence of clinical evidence of left atrial hypertension
- acute lung injury (ALI) considered to be present if PaO_2/FiO_2 is ≤ 300
- acute respiratory distress syndrome considered to be present if PaO_2/FiO_2 is ≤ 200.

In 2012, the Berlin Definition proposed 3 categories of ARDS based on degree of hypoxemia: mild (PaO_2/FiO_2 200-300 mm Hg), moderate (PaO_2/FiO_2 100-200 mm Hg), and severe (PaO_2/FiO_2 ≤ 100 mm Hg) and 4 ancillary variables for severe ARDS: radiographic severity, respiratory system compliance (≤ 40 mL/cm H_2O), positive end-expiratory pressure (≥ 10 cm H_2O), and corrected expired volume per minute (≥ 10 L/min) [3,4].

PATHOPHYSIOLOGY OF ALI/ARDS AS A RATIONALE FOR THE TREATMENT

ARDS is clinically characterized by acute lung injury, non-cardiogenic lung edema, and refractory hypoxemia. Diffuse alveolar damage with neutrophil infiltration, alveolar haemorrhage, and hyaline membrane formation are dominant histopathological features, fluently overlaping from acute (exudative) to fibroproliferative phase with various degrees of fibrosis, neovascularization and later resolution [5,6].

ALI/ARDS may result from direct lung injury (pneumonia, aspiration, near-drowning or toxic inhalation) or from indirect lung injury (sepsis, trauma, blood transfusion or pancreatitis). Variability in both pathology and patients is supposed to be responsible for contradictory results from many ARDS clinical trials, particularly in adult patients [7].

PHARMACOTHERAPY OF ALI/ARDS

In the pharmacotherapy of ARDS, several types of medicaments have been used (exogenous surfactant, corticosteroids, inhaled NO, inhaled prostacyclin, antioxidants, protease inhibitors etc. However, none of the used medicaments has proven to be commonly effective, although some of them may be beneficial in a subgroup of patients with specific causes of lung injury that might take them more responsive than others [8].

Several of the promisingly beneficial agents (corticosteroids, neutrophil elastase inhibitors, and anti-coagulant agents) were already discussed in the previous chapter by Hamada. Therefore, these medicaments are omitted in this chapter.

Exogenous Surfactant

In ARDS, dysfunction of pulmonary surfactant originates from quantitative and qualitative abnormalities of both phospholipids and proteins [7].

There are several pathomechanisms likely responsible for surfactant dysfunction in acute lung injury:

- reduced surfactant synthesis by directly or indirectly injured type II cells;
- functional inhibition of surfactant by plasma components, i.e. inhibition by competitive adsorption of plasma proteins and/or dysfunctional formation of surfactant film due to accumulation of plasma constituents;
- increased breakdown by activated oxidative, hydrolytic and proteolytic pathways;
- dilution of surfactant by lung edema fluid [7].

Once the endogenous surfactant becomes inactivated, some components of surfactant could be substituted by exogenous surfactant. Exogenous surfactant increases the pool of intraalveolar surfactant and enhances its production, and thereby increases the ratio surfactant/inhibitors [9]. Surfactant delivery increases the functional residual capacity, enlarges the diffusion area for gas exchange due to reduced atelectasis, and provides more homogenous distribution of ventilation. Local changes in the lung compliance and redistribution of the blood flow enhance oxygenation and decrease pulmonary vascular resistance [10]. Exogenous surfactant diminishes lung edema formation and inflammation and may reduce lung injury and occurence of complications [11,12].

Types, Dosage and Delivery of Exogenous Surfactants

Best preparation, optimal dose and timing of surfactant administration in ALI/ARDS is still discussed [13].

Evidence from randomized controlled trials in preterm infants with RDS indicates that treatment with natural surfactants results in faster weaning of supplemental oxygen and mean airway pressure, decreased duration of mechanical ventilation, and decreased mortality when compared to synthetic surfactants [14,15]. On the other hand, synthetic surfactants may be enriched with substances that improve spreading of surfactant in the alveoli and increase its resistance to inactivation, and their effects might be comparable with modified natural surfactants [14,16]. Despite several meta-analyses showed comparable results (i.e. death, chronic lung disease, clinical outcomes or incidence of RDS) between synthetic and natural (or animal-derived) surfactants [17], as well as between protein-containing synthetic surfactants and protein-free synthetic surfactant [18] for the prevention of RDS, there are no approved synthetic surfactants available for use in preterm infants.

Exogenous surfactant should be given at a sufficient dose, and surfactant administration should be repeated, if oxygenation remains compromised. For treatment of ALI/ARDS, dose of exogenous surfactant is calculated per body weight or body surface area. To achieve a comparable dose to 100 mg/kg b.w. in premature newborns, an adult of 70 kg b.w. requires about 7 g of exogenous surfactant [13]. In clinical studies, where exogenous surfactant was instilled to adults with ALI/ARDS, the dose ranged between 25 mg/kg [19] to 300 mg/kg [20]. However, an instilled dose of surfactant (100 mg/kg) is suspended in a total fluid volume of about 90-280 ml at currently used surfactants (of concentrations 25-80 mg/ml). Thus, in patients with edema and severe respiratory failure minimizing the instilled volume should be considered [13].

In ALI/ARDS, exogenous surfactant is primarily delivered intratracheally through an endotracheal tube [21,22,23], as it is done in premature infants. In older patients, surfactant may be instilled also via a bronchoscope [20,24]. Exogenous surfactant instilled into the large airways spreads and distributes to the lung periphery, as spreading towards the alveoli is promoted by surface tension gradients from the place of instillation with lower surface tension to peripheral regions with higher surface tension [13,25,26].

Distribution of instilled surfactant may be facilitated by specific strategies of mechanical ventilation, e.g. by high-frequency jet ventilation (HFJV) [27,28] or partial liquid ventilation [29,30]. Spreading of exogenous surfactant may be improved also by various formulation lowering the viscosity [13], e.g., by modifying the physical formulation by changes in dispersion methodology, ionic environment, or temperature [31], or by addition of polymers [32,33].

As „classical" surfactant administration requires endotracheal intubation and includes possible risks and complications of the administration, such as bradycardia, hypoxia, and hypotension, and noninvasive respiratory support is increasingly used, alternative forms of surfactant replacement therapy have been explored [34]. For instance, nebulization may effectively deliver surfactant of reduced required dose with less effects on blood pressure and cerebral blood flow [35,36,37]. However, the studies used different surfactants, different devices for nebulization and different methods including timing [34,38]. Thus, despite promising potential of this delivery, it has not yet been replicated in practice [39,40].

Effects of Surfactant Administration

The effects of surfactant delivery may be divided into pulmonary, cardiac, and radiologic. The immediate pulmonary effects include rapidly improved oxygenation associated with increasing functional residual capacity and increase in lung compliance. On the other hand, effects on pulmonary artery pressure and pulmonary blood flow may be different: no change or increase in the pulmonary blood flow. Radiologic image shows recruitment of lung volume and reduced atelectasis after surfactant treatment [34,41].

Complications of Surfactant Administration

Administration of surfactant may be associated with complications or adverse effects [10]. The most serious are hemodynamic changes, such as bradycardia, systemic hypotension, and changes of cerebral blood flow velocity [42,43,44,45,46]. Additionally, intratracheal administration of surfactant bolus may cause transient hypoxemia, drug reflux or the need to re-intubate [34].

Surfactant Treatment in the Neonates and Children

Replacement therapy with exogenous surfactant has become a routine intervention in the neonatal intensive care, as it prevents RDS in the premature neonates [14,47,48]. Nevertheless, administration of exogenous surfactant may be beneficial also in the term neonates with pneumonia or meconium aspiration syndrome. For instance, a single dose of exogenous surfactant significantly decreased oxygenation index, mean airway pressure and FiO_2, and increased PaO_2, arterial oxygen saturation and PaO_2/FiO_2 values in 18 term infants with respiratory failure within 6 hours after surfactant treatment [49].

In the multi-institutional, prospective, randomized, controlled, unblinded trial, intratracheal surfactant (calfactant) administration rapidly improved oxygenation, led to earlier extubation, and decreased requirement for intensive care in 42 children with acute hypoxemic respiratory failure [23]. In a later study included 153 infants, children, and adolescents (age 1 week to 21 years) with ALI, endotracheal instillation of calfactant acutely improved oxygenation and significantly decreased mortality although no significant decrease in the course of respiratory failure measured by duration of ventilator therapy, intensive care unit, or hospital stay was observed [22]. In a meta-analysis of 6 trials (314 patients), surfactant use decreased mortality, was associated with more ventilator-free days and reduced the duration of ventilation, while no serious adverse events were reported [21].

Surfactant Treatment in the Adult Patients

Despite surfactant therapy may be relevant for lung injury-related respiratory failure and ALI/ARDS in children, in adults it is still questionable [13]. In an international, multicenter, stratified, randomized, controlled trial, 418 adult patients with ALI/ARDS were included and received usual care either with or without instillation of exogenous natural porcine surfactant HL 10 as large boluses. Instillation of a large bolus of surfactant HL 10 did not improve 28-day outcome and showed a trend toward increased mortality and adverse effects [50]. In the meta-analysis of 9 randomized trials (2,575 patients), treatment with exogenous surfactant did not decrease mortality significantly. However, there was a significant effect on the change in the PaO_2/FiO_2 ratio in the first 24 hours, but this was lost by 120 hours. The duration of ventilation trended lower in surfactant-treated patients, but had a significantly higher risk of adverse effects [51]. In a recent meta-analysis of the use of exogenous surfactant for ARDS in adults, results of 7 trials (2,144 patients) showed that surfactant treatment was not associated with reduced mortality. Oxygenation, ventilation-free days, duration of ventilation, and APACHE II scores did not undergo pooled analysis due to insufficient data [52]. Therefore, at present exogenous surfactant cannot be considered an effective adjunctive therapy in adult patients with ALI/ARDS.

Pulmonary Vasodilators

In ARDS patients, inhaled vasodilators (inhaled NO or aerosolized prostacyclins) can improve hypoxemia, decrease pulmonary arterial pressure, and improve right-ventricular function and cardiac output without systemic hemodynamic effects. However, inhaled NO is expensive, potentially toxic, and requires complex technology for monitoring and administration. Inhaled prostacyclin may be a cheaper alternative to iNO, with comparable efficacy [53,54].

Inhaled Nitric Oxide (iNO)

Randomized controlled trial (108 children, median age 2.5 years) with acute hypoxemic respiratory failure demonstrated in the iNO group (10 ppm) an acute improvement in oxygenation during the first 12 hours [55].

However, other studies did not demonstrate clear improvement after iNO treatment. In a meta-analysis of 12 trials randomly assigning 1,237 patients (children and adults), no significant effect of NO on hospital mortality, duration of ventilation, or ventilator-free days was found. On the first days of treatment, NO increased PaO_2/FiO_2 ratio and decreased oxygenation index, but there was no effect on mean pulmonary arterial pressure. In addition, patients receiving NO had an increased risk of developing renal dysfunction [56]. Recent meta-analysis of 14 randomized controlled trials with a total of 1,303 participants (children and adults) showed no significant effect of iNO on overall mortality, duration of ventilation, ventilator-free days, or length of stay in the ICU and hospital. Authors presented a statistically significant but transient improvement in oxygenation in the first 24 hours. However, iNO appeared to increase the risk of renal impairment among adults, but not the risk of bleeding or methemoglobin or nitrogen dioxide formation [57]. Thus, despite iNO treatment can transiently improve oxygenation in patients with ALI or ARDS, it confers no mortality benefit and it cannot be recommended for routine use in these severely ill patients.

Aerosolized Prostacyclin

Prostacyclin (also called prostaglandin PGI_2) acts as an effective vasodilator and inhibits platelet activation. As a drug, it is also known as epoprostenol (EPO) [58].

In 9 adults with severe ARDS, inhaled nebulized prostacyclin caused significant dose-related improvement in oxygenation, with no significant dose effect on systemic or pulmonary arterial pressures, or on platelet function, as determined by platelet aggregation in response to challenge with adenosine diphosphate [59]. In a retrospective study, Camamo et al. [60] compared effects of inhaled alprostadil (PG E_1, n=17) or epoprostenol (PG I_2, n=10) on oxygenation in patients with ARDS, but found no significant changes in the PaO_2/FiO_2 ratio. In adult patients with pulmonary arterial hypertension, inhaled epoprostenol significantly decreased pulmonary pressures without lowering systemic blood pressure, showing minimal adverse events [53]. In a meta-analysis of only one randomized controlled study including 14 critically ill children with ALI or ARDS, aersosolized prostacyclin over less than 24 hours did not reduce overall mortality at 28 days compared with aerosolized saline. Despite the authors did not encounter any adverse events such as bleeding or organ dysfunction, they were also not able to assess the safety and efficacy of aerosolized prostacyclin for ALI and ARDS [61]. In a retrospective, single-center analysis of 105 adult, mechanically ventilated patients with ARDS receiving iNO or iEPO authors found no difference in the change in PaO_2/FiO_2 after 1 hour of therapy in the iNO and iEPO groups, respectively. No difference was observed in duration of therapy, mechanical ventilation, intensive care unit, and hospital lengths of stay comparing the iNO and iEPO groups. No adverse events were attributed to either therapy [62]. In other recent study, inhaled epoprostenol improved oxygenation by 10% or more in 62.5% (10/16) of patients with ARDS. However, hypotension was observed with a rate of 18.8% (3/16) [63].

Despite lack of sufficient data supporting the use of prostacycline as alternative of iNO, prostacycline is increasingly used because of the lower costs.

Beta-Agonists

β2 agonists may directly influence the neutrophil function and reduce the secretion of pro-inflammatory substances [64,65]. However, in patients with ARDS, intravenous salbutamol increased numbers of circulating neutrophils, but had no effect on alveolar neutrophils. Similarly, there was no effect on neutrophil chemotaxis, viability or apoptosis *in vitro* [66].

On the other hand, β2 agonists reduced the endothelial permability and enhanced the fluid clearance from the lungs [67,68]. β2 agonists may also stimulate the wound repair and spreading the cells in patients with ARDS [69]. These effects may be at least partially mediated through the up-regulation of matrix metalloproteinases (MMP), as salbutamol recently up-regulated MMP-9 *in vitro* by distal lung epithelial cells and *in vivo* in BAL fluid of ARDS patients [70].

Nevertheless, recent multicenter, placebo-controlled, randomised trial at 46 UK intensive-care units, which randomly assigned 162 patients to the salbutamol group and 164 to the placebo group showed that salbutamol increased 28-day mortality. Therefore, the routine use of β2 agonist treatment in patients with this disorder cannot be recommended [71].

Neuromuscular Agents

Neuromuscular blockade enables the lung ventilation in severely hypoxemic ARDS patients. Elimination of patient's effort via skeletal muscle inhibition may improve patient-ventilator synchrony, as evidenced by reduced airway pressures and better chest wall compliance [72]. Short-term paralysis can eliminate patient triggering, active expiratory muscle activity, and overventilation. These effects then may limit the regional overdistention (volutrauma) and cyclic alveolar collapse (atelectrauma), and lower the metabolism and overall ventilatory demand [73]. Finally, short-term paralysis may result in improved oxygenation and decreased ventilator-induced lung injury, but may also cause muscle weakness. Therefore the risk/benefit profile of these medications should be carefully considered.

In a recent multicenter double-blind trial, clinical outcomes after 2 days of therapy with neuromuscular blocking agent (cisatracurium besylate) in 178 patients with early, severe ARDS were evaluated. The authors found that early administration of a neuromuscular blocking agent improved the adjusted 90-day survival and increased the time off the ventilator without increasing muscle weakness in treated patients compared to patients treated with placebo [74].

Nevertheless, the potential role of neuromuscular blockers in ARDS patients with less severe disease is unclear [73].

Antioxidants

Regarding the role of reactive oxygen species in the pathogenesis of ALI/ARDS, antioxidants, e.g., N-acetylcysteine (NAC) can be useful in the treatment of ALI/ARDS.

NAC acts as a direct scavenger of hydrogen peroxide, hydroxol radicals, and hypochlorous acid. In addition, NAC is deacetylated to cysteine, a precursor of glutathione synthesis in cells, hence regulating the redox status in the cells [75]. By regulation the redox status in cells, NAC can interfere with several signaling pathways that play a role in regulation apoptosis, angiogenesis, cell growth, nuclear transcription etc. [76]. In addition, NAC can inhibit pulmonary fibrosis in ARDS, as it suppressed the effect of LPS on collagen content in the LPS-treated human embryonic lung fibroblasts [77]. Furthermore, NAC can reduce generation of pro-inflammatory substances (e.g. TNFα and IL-1) in the cells [78]. Alongside potent antioxidant and anti-inflammatory effects, NAC decreases viscosity and elasticity of mucus because of its ability to reduce disulphide bonds (Sadowska et al. 2006), what might be useful for clearance of the airways. In addition, NAC may partially prevent surfactant dysfunction as it might stimulate expression of surfactant protein (SP)-A mRNA [79]. Thanks to low bioavailability, NAC has rare adverse effects, such as anaphylaxis, tachycardia and hypotension, or reduced chemotaxis and increased cytotoxicity to polymorphonuclears, which are limited to very high concentrations [80].

In a rat model of LPS-induced ALI, systemic administration of NAC either before or after intratracheal instillation of LPS significantly attenuated the increase in lung permeability and reduced the lipid peroxidation. However, total and differential BAL cell counts and myeloperoxidase content were not affected by NAC pretreatment, indicating that attenuation of ALI by NAC was related to free radical scavenging and inhibition of the neutrophil oxidative burst, rather than by an effect on inflammatory cell migration. Although neutrophil influx was unaffected, neutrophil activation as assessed by surface CD11b expression and chemiluminescence was significantly down-regulated. Importantly, NAC administration up to 2 h after endotoxin challenge was still able to ameliorate LPS-induced lung injury [81]. In other study, NAC improved the LPS-induced hypotension and leukocytopenia, reduced the lung edema formation, exhaled NO, protein concentration in BAL fluid, decreased nitrate/nitrite, methyl guanidine, TNF-α and IL-1β, and improved lung pathology. During a 6 h observation period, pre-treatment with NAC significantly increased the survival rate [82]. Favourable results were observed also in the study by Choi et al. [83], where NAC treatment decreased lipid peroxidation, activity of myeloperoxidase and concentration of NF-κB, and reduced lung injury in a rat model of ALI, but had no effect on concentrations of TNF-α and IL-1β in BAL fluid.

Although results from animal models are almost positive, effects of NAC in clinical studies are little controversial. For instance, in a small group of ARDS patients, NAC increased extracellular total anti-oxidant power and total thiol molecules and also improved intracellular glutathione and the outcome of the patients [84]. Similarly, in patients presenting with mild-to-moderate ALI subsequent to a variety of underlying diseases, intravenous NAC treatment (n=32) during 72 h improved systemic oxygenation and reduced the need for ventilatory support, but did not reduce mortality vs. placebo group [85]. However, three years later, the same group of authors published that in a relatively small group of patients presenting with an established ARDS, intravenous NAC during 72 hours neither improved systemic oxygenation nor reduced the need for ventilatory support [86]. Similarly, Konrad et al. [87] showed that intravenous administration of 3 g NAC/day had no clinically relevant effect on glutathione levels, lipid peroxidation products, tracheobronchial mucus, and clinical condition in ventilated patients. In a recent meta-analysis of 41 fully published studies (2,768 patients with SIRS or sepsis), there was no effect of NAC on mortality, length of stay, duration of mechanical ventilation or incidence of new organ failure. Even, late NAC application was associated with cardiovascular instability [88]. Recently, long-time low-dose NAC treatment has increased expressions of proinflammatory cytokines through enhancement of kinase phosphorylation [89].

Statins

3-hydroxy-3-methylglutaryl coenzyme A (HMG CoA)-reductase inhibitors or statins are commonly used as lipid-lowering drugs for hypercholesterolemia. However, they have pleiotropic anti-inflammatory, immunomodulatory, antithrombotic, and antioxidant properties, and an ability to stimulate the resolution of inflammation found independently of the lipid-lowering properties [90].

For example, simvastatin attenuated vascular leak and inflammation in murine inflammatory lung injury [91], and improved survival in mice with sepsis [92]. Fluvastatin inhibited *in vivo* complement-dependent acute peritoneal inflammation [93].

Precise mechanisms of anti-inflammatory action of statins are not still completely elucidated. Statins decrease TLR-4 and 2 receptors expression and downstream signaling in human CD14+ monocytes [94,95]. Pretreatment of the cells with pravastatin or simvastatin reduced the LPS-induced transcription of NFkappaB and expression of IL-6 and TNF-alpha [96].

In addition, statins upregulate the anti-oxidant defense protein heme-oxygenase-1 [97,98], which may represent a novel target for statins in endothelial cells, beneficially influencing their function [99,100].

Anti-inflammatory potential of statins (e.g., lovastatin) can be also related to the ability to increase 15-epi-lipoxin A4 [101], which inhibits myeloperoxidase signaling, and thereby reduces production of oxidants, prevents delay in neutrophil apoptosis and prolongation of inflammation, and enhances resolution of ALI [102]. In addition, simvastatin may promote function of endothelial progenitor cells involved in endothelium repair of ALI [103].

Meta-analysis of total 20 studies demonstrated a protective effect of statins in patients with sepsis and/or other infections compared to placebo for various infection-related outcomes. However, the results are limited by the cohort design of the selected studies and the degree of heterogeneity among them, and further randomized trials are needed to validate the use of statins for sepsis and/or other infections [104].

In the randomized, double-blinded, placebo-controlled trial, 60 patients with ALI were treated with 80 mg simvastatin or placebo. Despite there was no difference in extravascular lung water and mortality, the simvastatin-treated group had improvements in nonpulmonary organ dysfunction, decreased IL-8 in BALF, and non-significantly improved oxygenation and respiratory mechanics. Simvastatin was well tolerated, with no increase in adverse events [105].

A recent meta-analysis also revealed a beneficial role of statins for the risk of development and mortality associated with community acquired pneumonia. However, the results constitute very low quality evidence as per the GRADE framework due to observational study design, heterogeneity and publication bias [106].

On the other hand, there was no evidence of a statin protective effect on clinical outcomes and only modest differences were found in circulating biomarkers in community-acquired pneumonia in a recent multicenter inception cohort study [107]. Similarly, meta-analysis by van den Hoek et al. [108] showed no effect of statins on the risk of infections or on infection related deaths.

In addition, statins have serious adverse effects, including liver test abnormalities and rises in the plasma levels of creatine kinase, which may limit their use in critically-ill patients [90].

NON-CONVENTIONAL APPROACHES (CELL AND/OR GENE THERAPY)

Despite wide range of therapeutic approaches used for ALI/ARDS, these were not able to significantly decrease the mortality [109]. Recently, on-going advances in the understanding of the pathophysiology of ALI/ARDS have revealed therapeutic targets for cell and/or gene therapy [110,111].

Cell-Based Approaches

Stem cells, i.e., mesenchymal stem/stromal cells (MSCs) and embryonic stem (ES) cells have clonogenic and self-renewing capabilities and differentiate into multiple cell lineages [111,112].

MSCs can modulate the immune response to reduce lung injury and endothelial damage, while maintaining host immune-competence and also facilitating lung regeneration and repair [113]. MSCs migrate to the termination of acute inflammation in the lungs and differentiate into organ-specific cells [114], but they also increase the production of growth factors that may mobilize endogenous stem cells [111]. Both increased mobilization of endogenous progenitor cells and alterations in the local cytokine milieu in favor of repair may contribute to the protective effects of stem cells [115]. In addition, MSCs may shift the balance from a pro-inflammatory to an anti-inflammatory response, e.g., enhancing production of IL-10 [116] and suppressing TNF-α and IL-1 [117].

Cell-based approaches have been explored in numerous preclinical studies [118]. Treatment with MSCs has improved survival both by systemic administration [115,119] and *via* intrapulmonary delivery [116,120] in various rodents models of ALI/ARDS. In mice with endotoxin-induced ALI, administering MSCs directly into the airspaces effectively reduced the pulmonary edema and mortality, decreased TNF-alpha and MIP-2 in the BALF and plasma, while increased the anti-inflammatory IL-10 [116]. In a murine model of sepsis, intravenous delivery of MSCs reduced mortality and organ injury, while this effect was mediated primarily by increased production of IL-10 by monocytes and alveolar macrophages in response to PG E2 released by the MSCs [121]. In a murine ALI model, systemic instillation of MSCs caused a substantial decrease in lung airspace inflammation and vascular leakage, characterized by significant reductions in protein content, differential neutrophil counts, and pro-inflammatory cytokine (TNF-α, IL-6, and MIP-2) concentrations in BALF. In addition, treated lungs showed preserved alveolar architecture, an absence of apoptosis, and minimal inflammatory cell infiltration [122]. Oropharyngeal aspiration of human MSCs (hMSCs) in LPS-injured mice significantly reduced the pulmonary edema, expression of pro-inflammatory cytokines, neutrophil counts and total protein in the BALF. In addition, the anti-inflammatory effects of hMSCs were not dependent on localization to the lung, as intraperitoneal administration of hMSCs also attenuated LPS-induced inflammation in the lung [123].

Similar anti-inflammatory properties have been demonstrated also after instillation of human and murine adipose-derived stem cells (ASCs) in LPS-induced model of ALI in mice. Both types of ASCs decreased leukocyte migration into the alveoli, total protein and albumin concentrations in BALF, and MPO activity after the induction of ALI following both therapies. Additionally, cell therapy with both cell types effectively suppressed the expression of pro-inflammatory cytokines and increased IL-10 [52].

Preclinical studies have shown that MSCs appear to restore epithelial and endothelial function, via both paracrine and cell contact dependent effects. Nowadays, intensive research in this field continues [118,124,125,126].

Gene-Based Therapy

Gene-based therapy involves the insertion of genes or smaller nucleic acid sequences into cells and tissues to replace the function of a defective gene, or to alter the production of a specific gene product, in order to treat a disorder [110]. Gene-based therapy seems to be suitable for ALI/ARDS, as ALI/ARDS is an acute but transient process requiring short-lived gene expression, obviating the need for repeated therapies and reducing the risk of an adverse immunological response. Distal lung epithelium is selectively accessible via the tracheal route of administration, while pulmonary circulation and pulmonary endothelium may be targeted following intravenous administration. In addition, it is possible to target also other cells participating in the pathogenesis of ALI/ARDS, such as leukocytes and fibroblasts, and to selectively influence different phases of the injury and repair process [110].

Gene-based therapy requires the delivery of genes or smaller nucleic acid sequences into the cell nucleus using a carrier or vector. The vector (viral or non-viral) enables the gene to overcome barriers to entry into the cell, and to make its way to the nucleus to be transcribed and translated itself or to modulate transcription and/or translation of other genes [110]. Delivery of both viral vectors and plasmid-based or non-viral vectors is a potentially powerful approach to treat ALI/ARDS, but there are still serious limitations for its use in clinical medicine [109]. Gene transfer into the parenchyma is limited by presence of several barriers: lung architecture, the innate immune system, and immune activation [127,128].

In viral vector-delivered gene therapy, the viral genome is modified. The parts necessary for viral replication are removed and this segment is replaced with the gene of interest coupled to a promoter that drives its expression [110]. In ALI/ARDS, adenoviruses, adeno-associated viruses, or retroviruses including lentiviruses may be used. For instance, adenovirus, the most widely used vector for lung gene therapy, provides high-efficiency transduction in dividing and non-dividing cells and high expression of delivery genes. However, inflammation, immunological responses, and non-specificity of cell targeting are associated with this vector [129,130].

Non-viral vectors may represent an alternative for gene therapy because of their ability to be repeatedly administered and their generally good safety profile with much less inflammation and immune responses [131]. Non-viral gene-based strategies are used to deliver smaller DNA/RNA molecules, including complexes with lipids or polymers. For example, plasmids complexed with cationic lipids as lipoplexes or with polymersas polyplexes may protect naked DNA from degradation [132,133]. However, use of non-viral gene therapy in the lung is limited by side effects of certain vectors and lower efficiency of gene transfer [109].

The recent preclinical studies have shown that gene therapy may prevent pulmonary edema by restoring alveolar epithelial function. Alveolar fluid clearance is driven by Na^+ transport entering the alveolar epithelial cell via epithelial Na^+ channel (ENaC) on the apical surface and then being pumped out by Na,K-ATPase within the basolateral surface into the interstitium and the pulmonary circulation [134]. Thus, overexpression of genes such as ENaC and Na,K-ATPase regulated by β-adrenergic receptor or cAMP can enhance alveolar active transport to clear pulmonary edema [135,136].

Gene therapy may also modulate the lung inflammation by several ways [109]. Gene delivery of various cytokines and chemokines attenuating inflammation, e.g., anti-inflammatory cytokines IL-10 [137], IFN protein-10 [138], IL-12 [139], and transforming growth factor beta-1 [140] improved survival and reduced inflammation. Other promising approach is delivery of heme oxygenase-1, an enzyme modulating neutrophil activity, infiltration into the lung and resolution of inflammation by influencing apoptosis and cytokine production [141]. Delivery of heme oxygenase-1 expressing adenoviruses into the lungs of various ALI models reduced inflammation and lung injury [142,143,144].

As mentioned above, gene therapy by viral and non-viral vectors have shown any benefits in various models of ALI/ARDS. Nowadays, search for improved viral and non-viral vectors and delivery systems that are non-toxic and non-inflammatory, enhanced gene expression strategies and therapeutic targeting, and improved cellular uptake of vector continues in the laboratory conditions before these approaches can move to clinical trials [109,110].

Cell-Delivered Gene Therapy

Cell-based gene therapy is a combination of cell and gene therapy. In this approach, vector cells, e.g. MSCs, fibroblasts, or endothelial progenitor cells, deliver genes to the lung, enhancing the therapeutic potential of these cells, or upregulate gene expression for anti-inflammatory mediators, such as endothelial NO synthase [145,146], inhibitory kappa B [147,148], keratinocyte growth factor [149], or angiopoietin-1 [150,151].

For instance, Mei et al. [150] compared effects of syngeneic MSCs with or without transfection with plasmid containing the human angiopoietin 1 gene (pANGPT1) delivered through the jugular vein of mice 30 min after intratracheal instillation of lipopolysaccharide (LPS) inducing lung injury. Treatment with MSCs alone significantly reduced LPS-induced acute pulmonary inflammation in mice, while administration of pANGPT1-transfected MSCs further improved both alveolar inflammation and permeability [150]. In transgenic mice, chronic eNOS overexpression inhibited the production of inflammatory chemokines and cytokines from VILI associated with neutrophil infiltration into the air space [146] and resulted in resistance to LPS-induced hypotension, lung injury, and death, associated with the reduced vascular reactivity to NO and the reduced anti-inflammatory effects of NO [145]. On the other hand, administration of keratinocyte-growth factor engineered MSCs not only improved pulmonary microvascular permeability and reduced total severity scores of lung injury, but also mediated a down-regulation of proinflammatory responses (reducing IL-1β and TNF-α) and an up-regulation of anti-inflammatory responses (increasing cytokine IL-10) [149].

Although the results indicate synergistic effects of both cell and gene approaches, further research including clinical trials is required to assess the safety and efficacy of the cell-based gene therapy in various lung pathologies.

CONCLUSION

Acute lung injury/acute respiratory distress syndrome is a serious clinical problem. Despite improved understanding of its complex pathophysiology, both morbidity and mortality remain high. Several novel strategies have been tested, whereas some of them (e.g. cell-based or gene-based therapies) seem to be very promising. Nevertheless, these approaches need to be evaluated thoroughly in the experimental and clinical conditions, before they can be recommended for treatment of ALI/ARDS.

REFERENCES

[1] Villar J. What is the acute respiratory distress syndrome? *Respir Care.* 2011 Oct;56(10):1539-45.

[2] Bernard GR, Artigas A, Brigham KL, Carlet J, Falke K, Hudson L, Lamy M, Legall JR, Morris A, Spragg R. The American-European Consensus Conference on ARDS. Definitions, mechanisms, relevant outcomes, and clinical trial The American-European Consensus Conference on ARDS. Definitions, mechanisms, relevant outcomes, and

clinical trial coordination. *Am J Respir Crit Care Med.* 1994 Mar;149(3 Pt 1):818-24.

[3] ARDS Definition Task Force, Ranieri VM, Rubenfeld GD, Thompson BT, Ferguson ND, Caldwell E, Fan E, Camporota L, Slutsky AS. Acute respiratory distress syndrome: the Berlin Definition. *JAMA.* 2012 Jun 20;307(23):2526-33.

[4] Fioretto JR, de Carvalho WB. Temporal evolution of acute respiratory distress syndrome definitions. *J Pediatr (Rio J).* 2013 Nov-Dec;89(6):523-30.

[5] Gattinoni L, Bombino M, Pelosi P, Lissoni A, Pesenti A, Fumagalli R, Tagliabue M. Lung structure and function in different stages of severe adult respiratory distress syndrome. *JAMA.* 1994 Jun 8;271(22):1772-9.

[6] Ware LB, Matthay MA. The acute respiratory distress syndrome. *N Engl J Med.* 2000;342(18):1334-49.

[7] Dushianthan A, Cusack R, Goss V, Postle AD, Grocott MP. Clinical review: Exogenous surfactant therapy for acute lung injury/acute respiratory distress syndrome - where do we go from here? *Crit Care.* 2012 Nov 22;16(6):238.

[8] Matthay MA, Zemans RL. The acute respiratory distress syndrome: pathogenesis and treatment. *Annu Rev Pathol.* 2011;6:147-63.

[9] Sun B, Curstedt T, Robertson B. Surfactant inhibition in experimental meconium aspiration. *Acta Paediatr* 1993; 82: 182-189.

[10] Hentschel R, Jorch G. Acute side effects of surfactant treatment. *J Perinat Med.* 2002;30(2):143-8.

[11] Hilgendorff A, Rawer D, Doerner M, Tutdibi E, Ebsen M, Schmidt R, Guenther A, Gortner L, Reiss I. Synthetic and natural surfactant differentially modulate inflammation after meconium aspiration. *Intensive Care Med* 2003; 29: 2247-2254.

[12] Bersani I, Kunzmann S, Speer CP. Immunomodulatory properties of surfactant preparations. *Expert Rev Anti Infect Ther.* 2013 Jan;11(1):99-110.

[13] Raghavendran K, Willson D, Notter RH. Surfactant therapy for acute lung injury and acute respiratory distress syndrome. *Crit Care Clin* 2011; 27: 525-559.

[14] Ramanathan R. Choosing a right surfactant for respiratory distress syndrome treatment. *Neonatology.* 2009;95(1):1-5.

[15] Ramanathan R1, Bhatia JJ, Sekar K, Ernst FR. Mortality in preterm infants with respiratory distress syndrome treated with poractant alfa,

calfactant or beractant: a retrospective study. *J Perinatol.* 2013 Feb;33(2):119-25.

[16] Moya F, Maturana A. Animal-derived surfactants versus past and current synthetic surfactants: current status. *Clin Perinatol.* 2007 Mar;34(1):145-77, viii.

[17] Pfister RH1, Soll RF, Wiswell T. Protein containing synthetic surfactant versus animal derived surfactant extract for the prevention and treatment of respiratory distress syndrome. *Cochrane Database Syst Rev.* 2007 Oct 17;(4):CD006069.

[18] Pfister RH, Soll R, Wiswell TE. Protein-containing synthetic surfactant versus protein-free synthetic surfactant for the prevention and treatment of respiratory distress syndrome. *Cochrane Database Syst Rev.* 2009 Oct 7;(4):CD006180.

[19] Spragg RG1, Lewis JF, Wurst W, Häfner D, Baughman RP, Wewers MD, Marsh JJ. Treatment of acute respiratory distress syndrome with recombinant surfactant protein C surfactant. *Am J Respir Crit Care Med.* 2003 Jun 1;167(11):1562-6.

[20] Walmrath D, Grimminger F, Pappert D, Knothe C, Obertacke U, Benzing A, Günther A, Schmehl T, Leuchte H, Seeger W. Bronchoscopic administration of bovine natural surfactant in ARDS and septic shock: impact on gas exchange and haemodynamics. *Eur Respir J.* 2002 May;19(5):805-10.

[21] Duffett M, Choong K, Ng V, Randolph A, Cook DJ. Surfactant therapy for acute respiratory failure in children: a systematic review and meta-analysis. *Crit Care.* 2007;11(3):R66.

[22] Willson DF, Thomas NJ, Markovitz BP, Bauman LA, DiCarlo JV, Pon S, Jacobs BR, Jefferson LS, Conaway MR, Egan EA; Pediatric Acute Lung Injury and Sepsis Investigators. Effect of exogenous surfactant (calfactant) in pediatric acute lung injury: a randomized controlled trial. *JAMA.* 2005 Jan 26;293(4):470-6.

[23] Willson DF, Zaritsky A, Bauman LA, Dockery K, James RL, Conrad D, Craft H, Novotny WE, Egan EA, Dalton H. Instillation of calf lung surfactant extract (calfactant) is beneficial in pediatric acute hypoxemic respiratory failure. Members of the Mid-Atlantic Pediatric Critical Care Network. *Crit Care Med.* 1999 Jan;27(1):188-95.

[24] Pallua N, Warbanow K, Noah EM, Machens HG, Poets C, Bernhard W, Berger A. Intrabronchial surfactant application in cases of inhalation injury: first results from patients with severe burns and ARDS. *Burns.* 1998 May;24(3):197-206.

[25] Davis JM, Russ GA, Metlay L, Dickerson B, Greenspan BS. Short-term distribution kinetics of intratracheally administered exogenous lung surfactant. *Pediatr Res* 1992a; 31: 445-450.
[26] Espinosa FF, Shapiro AH, Fredberg JJ, Kamm RD. Spreading of exogenous surfactant in an airway. *J Appl Physiol* (1985). 1993 Nov;75(5):2028-39.
[27] Davis JM, Richter SE, Kendig JW, Notter RH. High-frequency jet ventilation and surfactant treatment of newborns with severe respiratory failure. *Pediatr Pulmonol* 1992b; 13: 108-112.
[28] Calkovska A, Sevecova-Mokra D, Javorka K, Petraskova M, Adamicova K. Exogenous surfactant administration by asymmetric high-frequency jet ventilation in experimental respiratory distress syndrome. *Croat Med J.* 2005 Apr;46(2):209-17.
[29] Leach CL, Greenspan JS, Rubenstein SD, Shaffer TH, Wolfson MR, Jackson JC, DeLemos R, Fuhrman BP. Partial liquid ventilation with perflubron in premature infants with severe respiratory distress syndrome. The LiquiVent Study Group. *N Engl J Med.* 1996 Sep 12;335(11):761-7.
[30] Leach CL, Holm B, Morin FC 3rd, Fuhrman BP, Papo MC, Steinhorn D, Hernan LJ. Partial liquid ventilation in premature lambs with respiratory distress syndrome: efficacy and compatibility with exogenous surfactant. *J Pediatr.* 1995 Mar;126(3):412-20.
[31] King DM, Wang Z, Palmer HJ, Holm BA, Notter RH. Bulk shear viscosities of endogenous and exogenous lung surfactants. *Am J Physiol Lung Cell Mol Physiol.* 2002 Feb;282(2):L277-84.
[32] Lu KW, Pérez-Gil J, Taeusch H. Kinematic viscosity of therapeutic pulmonary surfactants with added polymers. *Biochim Biophys Acta.* 2009 Mar;1788(3):632-7.
[33] Calkovska A, Some M, Linderholm B, Curstedt T, Robertson B. Therapeutic effects of exogenous surfactant enriched with dextran in newborn rabbits with respiratory failure induced by airway instillation of albumin. *Pulm Pharmacol Ther.* 2008a;21(2):393-400.
[34] Gupta S, Donn SM. Novel approaches to surfactant administration. *Crit Care Res Pract.* 2012;2012:278483.
[35] Berggren E, Liljedahl M, Winbladh B, Andreasson B, Curstedt T, Robertson B, Schollin J. Pilot study of nebulized surfactant therapy for neonatal respiratory distress syndrome. *Acta Paediatr.* 2000 Apr;89(4):460-4.

[36] Mazela J, Merritt TA, Finer NN. Aerosolized surfactants. *Curr Opin Pediatr.* 2007 Apr;19(2):155-62.
[37] Schermuly RT, Günther A, Weissmann N, Ghofrani HA, Seeger W, Grimminger F, Walmrath D. Differential impact of ultrasonically nebulized versus tracheal-instilled surfactant on ventilation-perfusion (VA/Q) mismatch in a model of acute lung injury. *Am J Respir Crit Care Med.* 2000 Jan;161(1):152-9.
[38] Arzhavitina A, Steckel H. Surface active drugs significantly alter the drug output rate from medical nebulizers. *Int J Pharm.* 2010 Jan 15;384(1-2):128-36.
[39] Shah S. Exogenous surfactant: intubated present, nebulized future? *World J Pediatr.* 2011 Feb;7(1):11-5.
[40] Pillow JJ, Minocchieri S. Innovation in surfactant therapy II: surfactant administration by aerosolization. *Neonatology.* 2012;101(4):337-44.
[41] Dinger J, Töpfer A, Schaller P, Schwarze R. Functional residual capacity and compliance of the respiratory system after surfactant treatment in premature infants with severe respiratory distress syndrome. *Eur J Pediatr.* 2002 Sep;161(9):485-90.
[42] Wagner MH, Segerer H, Koch H, Scheid A, Obladen M. Circulatory changes after surfactant bolus instillation in lung-lavaged adult rabbits. *Exp Lung Res.* 1996 Nov-Dec;22(6):667-76.
[43] Moen A, Yu XQ, Almaas R, Curstedt T, Saugstad OD. Acute effects on systemic circulation after intratracheal instillation of Curosurf or Survanta in surfactant-depleted newborn piglets. *Acta Paediatr.* 1998 Mar;87(3):297-303.
[44] Nuntnarumit P, Bada HS, Yang W, Korones SB. Cerebral blood flow velocity changes after bovine natural surfactant instillation. *J Perinatol.* 2000 Jun;20(4):240-3.
[45] Kaiser JR, Gauss CH, Williams DK. Surfactant administration acutely affects cerebral and systemic hemodynamics and gas exchange in very-low-birth-weight infants. *J Pediatr.* 2004 Jun;144(6):809-14.
[46] Vitali F, Galletti S, Aceti A, Aquilano G, Fabi M, Balducci A, Faldella G. Pilot observational study on haemodynamic changes after surfactant administration in preterm newborns with respiratory distress syndrome. *Ital J Pediatr.* 2014 Mar 5;40(1):26.
[47] Ramanathan R. Surfactant therapy in preterm infants with respiratory distress syndrome and in near-term or term newborns with acute RDS. *J Perinatol.* 2006 May;26 Suppl 1:S51-6.

[48] Speer CP, Sweet DG, Halliday HL. Surfactant therapy: past, present and future. *Early Hum Dev.* 2013 Jun;89 Suppl 1:S22-4.

[49] Beken S, Turkyılmaz C, Koc E, Hirfanoglu IM, Altuntas N. The effects of surfactant on oxygenation in term infants with respiratory failure. *Iran J Pediatr.* 2013 Aug;23(4):477-80.

[50] Kesecioglu J, Beale R, Stewart TE, Findlay GP, Rouby JJ, Holzapfel L, Bruins P, Steenken EJ, Jeppesen OK, Lachmann B. Exogenous natural surfactant for treatment of acute lung injury and the acute respiratory distress syndrome. *Am J Respir Crit Care Med.* 2009 Nov 15;180(10):989-94.

[51] Meng H, Sun Y, Lu J, Fu S, Meng Z, Scott M, Li Q. Exogenous surfactant may improve oxygenation but not mortality in adult patients with acute lung injury/acute respiratory distress syndrome: a meta-analysis of 9 clinical trials. *J Cardiothorac Vasc Anesth.* 2012 Oct;26(5):849-56.

[52] Zhang S, Danchuk SD, Imhof KM, Semon JA, Scruggs BA, Bonvillain RW, Strong AL, Gimble JM, Betancourt AM, Sullivan DE, Bunnell BA. Comparison of the therapeutic effects of human and mouse adipose-derived stem cells in a murine model of lipopolysaccharide-induced acute lung injury. *Stem Cell Res Ther.* 2013 Jan 29;4(1):13.

[53] Buckley MS, Feldman JP. Inhaled epoprostenol for the treatment of pulmonary arterial hypertension in critically ill adults. *Pharmacotherapy.* 2010 Jul;30(7):728-40.

[54] Siobal MS1, Hess DR. Are inhaled vasodilators useful in acute lung injury and acute respiratory distress syndrome? *Respir Care.* 2010 Feb;55(2):144-57; discussion 157-61.

[55] Dobyns EL, Cornfield DN, Anas NG, Fortenberry JD, Tasker RC, Lynch A, Liu P, Eells PL, Griebel J, Baier M, Kinsella JP, Abman SH. Multicenter randomized controlled trial of the effects of inhaled nitric oxide therapy on gas exchange in children with acute hypoxemic respiratory failure. *J Pediatr.* 1999 Apr;134(4):406-12.

[56] Adhikari NK1, Burns KE, Friedrich JO, Granton JT, Cook DJ, Meade MO. Effect of nitric oxide on oxygenation and mortality in acute lung injury: systematic review and meta-analysis. *BMJ.* 2007 Apr 14;334(7597):779.

[57] Afshari A, Brok J, Møller AM, Wetterslev J. Inhaled nitric oxide for acute respiratory distress syndrome (ARDS) and acute lung injury in children and adults. *Cochrane Database Syst Rev.* 2010a Jul 7;(7):CD002787.

[58] Siobal M. Aerosolized prostacyclins. *Respir Care.* 2004 Jun;49(6):640-52.
[59] van Heerden PV, Barden A, Michalopoulos N, Bulsara MK, Roberts BL. Dose-response to inhaled aerosolized prostacyclin for hypoxemia due to ARDS. *Chest. 2000* Mar;117(3):819-27.
[60] Camamo JM1, McCoy RH, Erstad BL. Retrospective evaluation of inhaled prostaglandins in patients with acute respiratory distress syndrome. *Pharmacotherapy.* 2005 Feb;25(2):184-90.
[61] Afshari A, Brok J, Møller AM, Wetterslev J. Aerosolized prostacyclin for acute lung injury (ALI) and acute respiratory distress syndrome (ARDS*). Cochrane Database Syst Rev.* 2010b Aug 4;(8):CD007733.
[62] Torbic H, Szumita PM, Anger KE, Nuccio P, LaGambina S, Weinhouse G. Inhaled epoprostenol vs inhaled nitric oxide for refractory hypoxemia in critically ill patients. *J Crit Care.* 2013 Oct;28(5):844-8.
[63] Dunkley KA, Louzon PR, Lee J, Vu S. Efficacy, safety, and medication errors associated with the use of inhaled epoprostenol for adults with acute respiratory distress syndrome: a pilot study. *Ann Pharmacother.* 2013 Jun;47(6):790-6.
[64] Hanania NA, Moore RH. Anti-inflammatory activities of beta2-agonists. *Curr Drug Targets Inflamm Allergy.* 2004 Sep;3(3):271-7.
[65] Theron AJ, Steel HC, Tintinger GR, Feldman C, Anderson R. Can the anti-inflammatory activities of β2-agonists be harnessed in the clinical setting? *Drug Des Devel Ther.* 2013 Nov 22;7:1387-98.
[66] Perkins GD, Nathani N, McAuley DF, Gao F, Thickett DR. In vitro and in vivo effects of salbutamol on neutrophil function in acute lung injury. *Thorax.* 2007 Jan;62(1):36-42.
[67] Sakuma T, Folkesson HG, Suzuki S, Okaniwa G, Fujimura S, Matthay MA. Beta-adrenergic agonist stimulated alveolar fluid clearance in ex vivo human and rat lungs. *Am J Respir Crit Care Med.* 1997 Feb;155(2):506-12.
[68] McAuley DF, Frank JA, Fang X, Matthay MA. Clinically relevant concentrations of beta2-adrenergic agonists stimulate maximal cyclic adenosine monophosphate-dependent airspace fluid clearance and decrease pulmonary edema in experimental acid-induced lung injury. *Crit Care Med.* 2004 Jul;32(7):1470-6.
[69] Perkins GD, Gao F, Thickett DR. In vivo and in vitro effects of salbutamol on alveolar epithelial repair in acute lung injury. *Thorax.* 2008 Mar;63(3):215-20.

[70] O'Kane CM, McKeown SW, Perkins GD, Bassford CR, Gao F, Thickett DR, McAuley DF. Salbutamol up-regulates matrix metalloproteinase-9 in the alveolar space in the acute respiratory distress syndrome. *Crit Care Med.* 2009 Jul;37(7):2242-9.

[71] Gao Smith F, Perkins GD, Gates S, Young D, McAuley DF, Tunnicliffe W, Khan Z, Lamb SE; BALTI-2 study investigators. Effect of intravenous β-2 agonist treatment on clinical outcomes in acute respiratory distress syndrome (BALTI-2): a multicentre, randomised controlled trial. *Lancet.* 2012 Jan 21;379(9812):229-35.

[72] Boyle AJ, Mac Sweeney R, McAuley DF. Pharmacological treatments in ARDS; a state-of-the-art update. *BMC Med.* 2013 Aug 20;11:166. doi: 10.1186/1741-7015-11-166.

[73] Donahoe M. Acute respiratory distress syndrome: A clinical review. *Pulm Circ.* 2011 Apr-Jun;1(2):192-211.

[74] Papazian L, Forel JM, Gacouin A, Penot-Ragon C, Perrin G, Loundou A, Jaber S, Arnal JM, Perez D, Seghboyan JM, Constantin JM, Courant P, Lefrant JY, Guérin C, Prat G, Morange S, Roch A; ACURASYS Study Investigators. Neuromuscular blockers in early acute respiratory distress syndrome. *N Engl J Med.* 2010 Sep 16;363(12):1107-16.

[75] Aruoma OI, Halliwell B, Hoey BM, Butler J. The antioxidant action of N-acetylcysteine: its reaction with hydrogen peroxide, hydroxyl radical, superoxide, and hypochlorous acid. *Free Radic Biol Med* 1989; 6: 593-597.

[76] Gillissen A. [Anti-inflammatory efficacy of N-acetylcysteine and therapeutic usefulness]. [Article in German] *Pneumologie.* 2011 Sep;65(9):549-57.

[77] Li XF, Ouyang B, Wu JF, Chen J, Guan XD. [N-acetylcysteine (NAC) inhibited pulmonary fibrosis in acute respiratory distress syndrome (ARDS)]. [Article in Chinese] *Zhongguo Wei Zhong Bing Ji Jiu Yi Xue.* 2011 Oct;23(10):599-601.

[78] Haddad JJ. A redox microenvironment is essential for MAPK-dependent secretion of pro-inflammatory cytokines: modulation by glutathione (GSH/GSSG) biosynthesis and equilibrium in the alveolar epithelium. *Cell Immunol* 2011; 270: 53-61.

[79] Fu Z, Yang Z, Li A. The effects of NAC on the expression and activity of SPA in rats inflicted by smoke inhalation injury. *Zhonghua Shao Shang Za Zhi* 2000; 16: 173-176. (Abstract).

[80] Gillissen A, Nowak. Characterization of N-acetylcysteine and ambroxol in anti-oxidant therapy. *Respir Med* 1998; 92: 609-623.

[81] Davreux CJ, Soric I, Nathens AB, Watson RW, McGilvray ID, Suntres ZE, Shek PN, Rotstein OD. N-acetyl cysteine attenuates acute lung injury in the rat. *Shock.* 1997 Dec;8(6):432-8.
[82] Kao SJ, Wang D, Lin HI, Chen HI. N-acetylcysteine abrogates acute lung injury induced by endotoxin. *Clin Exp Pharmacol Physiol.* 2006 Jan-Feb;33(1-2):33-40.
[83] Choi JS, Lee HS, Seo KH, Na JO, Kim YH, Uh ST, Park CS, Oh MH, Lee SH, Kim YT. The effect of post-treatment N-acetylcysteine in LPS-induced acute lung injury of rats. *Tuberc Respir Dis (Seoul).* 2012 Jul;73(1):22-31.
[84] Soltan-Sharifi MS, Mojtahedzadeh M, Najafi A, Reza Khajavi M, Reza Rouini M, Moradi M, Mohammadirad A, Abdollahi M. Improvement by N-acetylcysteine of acute respiratory distress syndrome through increasing intracellular glutathione, and extracellular thiol molecules and anti-oxidant power: evidence for underlying toxicological mechanisms. *Hum Exp Toxicol.* 2007 Sep;26(9):697-703.
[85] Suter PM, Domenighetti G, Schaller MD, Laverrière MC, Ritz R, Perret C. N-acetylcysteine enhances recovery from acute lung injury in man. A randomized, double-blind, placebo-controlled clinical study. *Chest.* 1994 Jan;105(1):190-4.
[86] Domenighetti G, Suter PM, Schaller MD, Ritz R, Perret C. Treatment with N-acetylcysteine during acute respiratory distress syndrome: a randomized, double-blind, placebo-controlled clinical study. *J Crit Care.* 1997 Dec;12(4):177-82.
[87] Konrad F, Schoenberg MH, Wiedmann H, Kilian J, Georgieff M. [The application of n-acetylcysteine as an antioxidant and mucolytic in mechanical ventilation in intensive care patients. A prospective, randomized, placebo-controlled, double-blind study]. [Article in German] *Anaesthesist.* 1995 Sep;44(9):651-8.
[88] Szakmany T, Hauser B, Radermacher P. N-acetylcysteine for sepsis and systemic inflammatory response in adults. *Cochrane Database Syst Rev.* 2012 Sep 12;9:CD006616.
[89] Ohnishi T, Bandow K, Kakimoto K, Kusuyama J, Matsuguchi T. Long-time treatment by low-dose N-acetyl-L-cysteine enhances proinflammatory cytokine expressions in LPS-stimulated macrophages. *PLoS One.* 2014 Feb 4;9(2):e87229.
[90] De Loecker I, Preiser JC. Statins in the critically ill. *Ann Intensive Care.* 2012 Jun 18;2(1):19.

[91] Jacobson JR, Barnard JW, Grigoryev DN, Ma SF, Tuder RM, Garcia JG. Simvastatin attenuates vascular leak and inflammation in murine inflammatory lung injury. *Am J Physiol Lung Cell Mol Physiol.* 2005 Jun;288(6):L1026-32.
[92] Merx MW, Liehn EA, Graf J, van de Sandt A, Schaltenbrand M, Schrader J, Hanrath P, Weber C. Statin treatment after onset of sepsis in a murine model improves survival. *Circulation.* 2005 Jul 5;112(1):117-24.
[93] Fischetti F, Carretta R, Borotto G, Durigutto P, Bulla R, Meroni PL, Tedesco F. Fluvastatin treatment inhibits leucocyte adhesion and extravasation in models of complement-mediated acute inflammation. *Clin Exp Immunol.* 2004 Feb;135(2):186-93.
[94] Methe H, Kim JO, Kofler S, Nabauer M, Weis M. Statins decrease Toll-like receptor 4 expression and downstream signaling in human CD14+ monocytes. *Arterioscler Thromb Vasc Biol.* 2005 Jul;25(7):1439-45.
[95] Niessner A, Steiner S, Speidl WS, Pleiner J, Seidinger D, Maurer G, Goronzy JJ, Weyand CM, Kopp CW, Huber K, Wolzt M, Wojta J. Simvastatin suppresses endotoxin-induced upregulation of toll-like receptors 4 and 2 in vivo. *Atherosclerosis.* 2006 Dec;189(2):408-13. Epub 2006 Jan 26.
[96] Hodgkinson CP, Ye S. Statins inhibit Toll-like receptor 4-mediated lipopolysaccharide signalling and cytokine expression. *Pharmacogenet Genomics.* 2008;18:803–813.
[97] Grosser N, Erdmann K, Hemmerle A, Berndt G, Hinkelmann U, Smith G, Schröder H. Rosuvastatin upregulates the antioxidant defense protein heme oxygenase-1. *Biochem Biophys Res Commun.* 2004a;325(3):871-6.
[98] Lee TS, Chang CC, Zhu Y, Shyy JY. Simvastatin induces heme oxygenase-1: a novel mechanism of vessel protection. *Circulation.* 2004;110:1296–1302.
[99] Wassmann S, Laufs U, Müller K, Konkol C, Ahlbory K, Bäumer AT, Linz W, Böhm M, Nickenig G. Cellular antioxidant effects of atorvastatin in vitro and in vivo. *Arterioscler Thromb Vasc Biol.* 2002 Feb 1;22(2):300-5.
[100] Grosser N, Hemmerle A, Berndt G, Erdmann K, Hinkelmann U, Schürger S, Wijayanti N, Immenschuh S, Schröder H. The antioxidant defense protein heme oxygenase 1 is a novel target for statins in endothelial cells. *Free Radic Biol Med.* 2004b;37(12):2064-71.

[101] Planagumà A, Pfeffer MA, Rubin G, Croze R, Uddin M, Serhan CN, Levy BD. Lovastatin decreases acute mucosal inflammation via 15-epi-lipoxin A4. *Mucosal Immunol.* 2010 May;3(3):270-9.

[102] El Kebir D, József L, Pan W, Wang L, Petasis NA, Serhan CN, Filep JG. 15-epi-lipoxin A4 inhibits myeloperoxidase signaling and enhances resolution of acute lung injury. *Am J Respir Crit Care Med.* 2009 Aug 15;180(4):311-9.

[103] Li H, Qiang Y, Wang L, Wang G, Yi J, Jing H, Wu H. Repair of lipopolysaccharide-induced acute lung injury in mice by endothelial progenitor cells, alone and in combination with simvastatin. *Chest.* 2013 Sep;144(3):876-86.

[104] Janda S, Young A, Fitzgerald JM, Etminan M, Swiston J. The effect of statins on mortality from severe infections and sepsis: a systematic review and meta-analysis. *J Crit Care.* 2010 Dec;25(4):656.e7-22.

[105] Craig TR, Duffy MJ, Shyamsundar M, McDowell C, O'Kane CM, Elborn JS, McAuley DF. A randomized clinical trial of hydroxymethylglutaryl- coenzyme a reductase inhibition for acute lung injury (The HARP Study). *Am J Respir Crit Care Med.* 2011 Mar 1;183(5):620-6.

[106] Khan AR, Riaz M, Bin Abdulhak AA, Al-Tannir MA, Garbati MA, Erwin PJ, Baddour LM, Tleyjeh IM. The role of statins in prevention and treatment of community acquired pneumonia: a systematic review and meta-analysis. *PLoS One.* 2013;8(1):e52929.

[107] Yende S, Milbrandt EB, Kellum JA, Kong L, Delude RL, Weissfeld LA, Angus DC. Understanding the potential role of statins in pneumonia and sepsis. *Crit Care Med.* 2011 Aug;39(8):1871-8.

[108] van den Hoek HL, Bos WJ, de Boer A, van de Garde EM. Statins and prevention of infections: systematic review and meta-analysis of data from large randomised placebo controlled trials. *BMJ.* 2011 Nov 29;343:d7281.

[109] Lin X, Dean DA. Gene therapy for ALI/ARDS. *Crit Care Clin.* 2011 Jul;27(3):705-18.

[110] Devaney J, Contreras M, Laffey JG. Clinical review: gene-based therapies for ALI/ARDS: where are we now? *Crit Care.* 2011;15(3):224.

[111] Zhu YG1, Qu JM, Zhang J, Jiang HN, Xu JF. Novel interventional approaches for ALI/ARDS: cell-based gene therapy. *Mediators Inflamm.* 2011;2011:560194.

[112] Weissman IL. Stem cells: units of development, units of regeneration, and units in evolution. *Cell.* 2000 Jan 7;100(1):157-68.
[113] Hayes M, Curley G, Ansari B, Laffey JG. Clinical review: Stem cell therapies for acute lung injury/acute respiratory distress syndrome - hope or hype? *Crit Care.* 2012 Dec 12;16(2):205.
[114] Abkowitz JL, Robinson AE, Kale S, Long MW, Chen J. Mobilization of hematopoietic stem cells during homeostasis and after cytokine exposure. *Blood.* 2003 Aug 15;102(4):1249-53.
[115] Rojas M, Xu J, Woods CR, Mora AL, Spears W, Roman J, Brigham KL. Bone marrow-derived mesenchymal stem cells in repair of the injured lung. *Am J Respir Cell Mol Biol.* 2005 Aug;33(2):145-52.
[116] Gupta N, Su X, Popov B, Lee JW, Serikov V, Matthay MA. Intrapulmonary delivery of bone marrow-derived mesenchymal stem cells improves survival and attenuates endotoxin-induced acute lung injury in mice. *J Immunol.* 2007 Aug 1;179(3):1855-63.
[117] Ortiz LA, Dutreil M, Fattman C, Pandey AC, Torres G, Go K, Phinney DG. Interleukin 1 receptor antagonist mediates the antiinflammatory and antifibrotic effect of mesenchymal stem cells during lung injury. *Proc Natl Acad Sci U S A.* 2007 Jun 26;104(26):11002-7.
[118] Maron-Gutierrez T, Laffey JG, Pelosi P, Rocco PR. Cell-based therapies for the acute respiratory distress syndrome. *Curr Opin Crit Care.* 2014 Feb;20(1):122-31.
[119] Ortiz LA, Gambelli F, McBride C, Gaupp D, Baddoo M, Kaminski N, Phinney DG. Mesenchymal stem cell engraftment in lung is enhanced in response to bleomycin exposure and ameliorates its fibrotic effects. *Proc Natl Acad Sci U S A.* 2003 Jul 8;100(14):8407-11.
[120] Xu J, Woods CR, Mora AL, Joodi R, Brigham KL, Iyer S, Rojas M. Prevention of endotoxin-induced systemic response by bone marrow-derived mesenchymal stem cells in mice. *Am J Physiol Lung Cell Mol Physiol.* 2007 Jul;293(1):L131-41.
[121] Németh K, Leelahavanichkul A, Yuen PS, Mayer B, Parmelee A, Doi K, Robey PG, Leelahavanichkul K, Koller BH, Brown JM, Hu X, Jelinek I, Star RA, Mezey E. Bone marrow stromal cells attenuate sepsis via prostaglandin E(2)-dependent reprogramming of host macrophages to increase their interleukin-10 production. *Nat Med.* 2009 Jan;15(1):42-9.
[122] Martínez-González I, Roca O, Masclans JR, Moreno R, Salcedo MT, Baekelandt V, Cruz MJ, Rello J, Aran JM. Human mesenchymal stem cells overexpressing the IL-33 antagonist soluble IL-1 receptor-like-1

attenuate endotoxin-induced acute lung injury. *Am J Respir Cell Mol Biol.* 2013 Oct;49(4):552-62.

[123] Danchuk S, Ylostalo JH, Hossain F, Sorge R, Ramsey A, Bonvillain RW, Lasky JA, Bunnell BA, Welsh DA, Prockop DJ, Sullivan DE. Human multipotent stromal cells attenuate lipopolysaccharide-induced acute lung injury in mice via secretion of tumor necrosis factor-α-induced protein 6. *Stem Cell Res Ther.* 2011 May 13;2(3):27.

[124] Cárdenes N, Cáceres E, Romagnoli M, Rojas M. Mesenchymal stem cells: a promising therapy for the acute respiratory distress syndrome. *Respiration.* 2013;85(4):267-78.

[125] Curley GF, Laffey JG. Cell therapy demonstrates promise for acute respiratory distress syndrome - but which cell is best? *Stem Cell Res Ther.* 2013 Mar 22;4(2):29.

[126] Masterson C, Jerkic M, Curley GF, Laffey JG. Mesenchymal Stromal cell therapies - potential and pitfalls for ARDS. *Minerva Anestesiol.* 2014 Feb 4. [Epub ahead of print]

[127] Kolb M, Martin G, Medina M, Ask K, Gauldie J. Gene therapy for pulmonary diseases. *Chest.* 2006 Sep;130(3):879-84.

[128] Geiger J, Aneja MK, Rudolph C. Vectors for pulmonary gene therapy. *Int J Pharm.* 2010 May 5;390(1):84-8.

[129] Muruve DA. The innate immune response to adenovirus vectors. *Hum Gene Ther.* 2004 Dec;15(12):1157-66.

[130] Kushwah R, Cao H, Hu J. Potential of helper-dependent adenoviral vectors in modulating airway innate immunity. *Cell Mol Immunol.* 2007 Apr;4(2):81-9.

[131] Lam AP, Dean DA. Progress and prospects: nuclear import of nonviral vectors. *Gene Ther.* 2010 Apr;17(4):439-47.

[132] De Smedt SC, Demeester J, Hennink WE. Cationic polymer based gene delivery systems. *Pharm Res.* 2000 Feb;17(2):113-26.

[133] Godbey WT1, Barry MA, Saggau P, Wu KK, Mikos AG. Poly(ethylenimine)-mediated transfection: a new paradigm for gene delivery. *J Biomed Mater Res.* 2000 Sep 5;51(3):321-8.

[134] Budinger GR, Sznajder JI. The alveolar-epithelial barrier: a target for potential therapy. *Clin Chest Med.* 2006 Dec;27(4):655-69.

[135] Factor P, Dumasius V, Saldias F, Brown LA, Sznajder JI. Adenovirus-mediated transfer of an Na+/K+-ATPase beta1 subunit gene improves alveolar fluid clearance and survival in hyperoxic rats. *Hum Gene Ther.* 2000 Nov 1;11(16):2231-42.

[136] Machado-Aranda D1, Adir Y, Young JL, Briva A, Budinger GR, Yeldandi AV, Sznajder JI, Dean DA. Gene transfer of the Na+,K+-ATPase beta1 subunit using electroporation increases lung liquid clearance. *Am J Respir Crit Care Med.* 2005 Feb 1;171(3):204-11.

[137] Buff SM1, Yu H, McCall JN, Caldwell SM, Ferkol TW, Flotte TR, Virella-Lowell IL. IL-10 delivery by AAV5 vector attenuates inflammation in mice with Pseudomonas pneumonia. *Gene Ther.* 2010 May;17(5):567-76.

[138] McAllister F1, Ruan S, Steele C, Zheng M, McKinley L, Ulrich L, Marrero L, Shellito JE, Kolls JK. CXCR3 and IFN protein-10 in Pneumocystis pneumonia. *J Immunol.* 2006 Aug 1;177(3):1846-54.

[139] Ruan S1, McKinley L, Zheng M, Rudner X, D'Souza A, Kolls JK, Shellito JE. Interleukin-12 and host defense against murine Pneumocystis pneumonia. *Infect Immun.* 2008 May;76(5):2130-7.

[140] Mora BN, Boasquevisque CH, Boglione M, Ritter JM, Scheule RK, Yew NS, Debruyne L, Qin L, Bromberg JS, Patterson GA. Transforming growth factor-beta1 gene transfer ameliorates acute lung allograft rejection. *J Thorac Cardiovasc Surg.* 2000 May;119(5):913-20.

[141] Fredenburgh LE, Perrella MA, Mitsialis SA. The role of heme oxygenase-1 in pulmonary disease. *Am J Respir Cell Mol Biol.* 2007 Feb;36(2):158-65.

[142] Hashiba T, Suzuki M, Nagashima Y, Suzuki S, Inoue S, Tsuburai T, Matsuse T, Ishigatubo Y. Adenovirus-mediated transfer of heme oxygenase-1 cDNA attenuates severe lung injury induced by the influenza virus in mice. *Gene Ther.* 2001 Oct;8(19):1499-507.

[143] Inoue S, Suzuki M, Nagashima Y, Suzuki S, Hashiba T, Tsuburai T, Ikehara K, Matsuse T, Ishigatsubo Y. Transfer of heme oxygenase 1 cDNA by a replication-deficient adenovirus enhances interleukin 10 production from alveolar macrophages that attenuates lipopolysaccharide-induced acute lung injury in mice. *Hum Gene Ther.* 2001 May 20;12(8):967-79.

[144] Tsuburai T, Kaneko T, Nagashima Y, Ueda A, Tagawa A, Shinohara T, Ishigatsubo Y. Pseudomonas aeruginosa-induced neutrophilic lung inflammation is attenuated by adenovirus-mediated transfer of the heme oxygenase 1 cDNA in mice. *Hum Gene Ther.* 2004 Mar;15(3):273-85.

[145] Yamashita T, Kawashima S, Ohashi Y, Ozaki M, Ueyama T, Ishida T, Inoue N, Hirata K, Akita H, Yokoyama M. Resistance to endotoxin

shock in transgenic mice overexpressing endothelial nitric oxide synthase. *Circulation.* 2000 Feb 29;101(8):931-7.

[146] Takenaka K, Nishimura Y, Nishiuma T, Sakashita A, Yamashita T, Kobayashi K, Satouchi M, Ishida T, Kawashima S, Yokoyama M. Ventilator-induced lung injury is reduced in transgenic mice that overexpress endothelial nitric oxide synthase. *Am J Physiol Lung Cell Mol Physiol.* 2006 Jun;290(6):L1078-86.

[147] Wrighton CJ, Hofer-Warbinek R, Moll T, Eytner R, Bach FH, de Martin R. Inhibition of endothelial cell activation by adenovirus-mediated expression of I kappa B alpha, an inhibitor of the transcription factor NF-kappa B. *J Exp Med.* 1996 Mar 1;183(3):1013-22.

[148] Makarov SS, Johnston WN, Olsen JC, Watson JM, Mondal K, Rinehart C, Haskill JS. NF-kappa B as a target for anti-inflammatory gene therapy: suppression of inflammatory responses in monocytic and stromal cells by stable gene transfer of I kappa B alpha cDNA. *Gene Ther.* 1997 Aug;4(8):846-52.

[149] Chen J, Li C, Gao X, Li C, Liang Z, Yu L, Li Y, Xiao X, Chen L. Keratinocyte Growth Factor Gene Delivery via Mesenchymal Stem Cells Protects against Lipopolysaccharide-Induced Acute Lung Injury in Mice. *PLoS One.* 2013 Dec 18;8(12):e83303.

[150] Mei SH, McCarter SD, Deng Y, Parker CH, Liles WC, Stewart DJ. Prevention of LPS-induced acute lung injury in mice by mesenchymal stem cells overexpressing angiopoietin 1. *PLoS Med.* 2007 Sep;4(9):e269.

[151] McCarter SD, Mei SH, Lai PF, Zhang QW, Parker CH, Suen RS, Hood RD, Zhao YD, Deng Y, Han RN, Dumont DJ, Stewart DJ. Cell-based angiopoietin-1 gene therapy for acute lung injury. *Am J Respir Crit Care Med.* 2007 May 15;175(10):1014-26.

Reviewed by:
Prof. Andrea Calkovska, M.D., Ph.D., Department of Physiology, Jessenius Faculty of Medicine, Comenius University, Martin, Slovakia Assoc. Prof. Jana Plevkova, M.D., Ph.D., Department of Pathological Physiology, Jessenius Faculty of Medicine, Comenius University, Martin, Slovakia.

EDITOR CONTACT INFORMATION

Daniela Mokrá, M.D., Ph.D.
Comenius University in Bratislava
Jessenius Faculty of Medicine in Martin
Mala Hora 4
SK-036 01 Martin
Slovakia, EU
Tel: +421 43 2633454
E-mail: mokra@jfmed.uniba.sk

INDEX

A

Abraham, 44
accounting, 21
acid, 40, 43, 54, 59, 67, 68
active transport, 60
acute lung injury, vii, viii, 1, 2, 10, 17, 18, 19, 29, 30, 31, 33, 34, 35, 36, 42, 43, 44, 45, 46, 47, 62, 63, 65, 66, 67, 69, 71, 72, 73, 74, 75
acute respiratory distress syndrome, viii, 2, 10, 20, 31, 37, 42, 43, 44, 46, 61, 62, 63, 66, 67, 68, 69, 72, 73
adaptation, 4
adenosine, 52, 67
adenovirus, 59, 73, 74, 75
adhesion(s), 22, 23, 27, 28, 30, 31, 35, 38, 70
adipose, 58, 66
adjunctive therapy, 51
adolescents, 50
adsorption, 48
adult respiratory distress syndrome, 38, 42, 43, 62
adults, 42, 44, 49, 51, 52, 66, 67, 69
adverse effects, 50, 51, 54, 57
adverse event, 18, 39, 40, 50, 52, 57
age, 50, 51
aggregation, 52
agonist, viii, 20, 53, 67, 68

airways, 49, 54
albumin, 58, 64
ALI, viii, 19, 20, 21, 22, 23, 24, 26, 27, 28, 29, 30, 38, 39, 40, 41, 45, 46, 47, 48, 49, 50, 51, 52, 54, 55, 56, 57, 58, 59, 60, 61, 67, 71
alters, 28, 35
alveolar damage, viii, 20, 45, 47
alveolar epithelium, viii, 19, 20, 21, 27, 68
alveolar liquid clearance capacity, viii, 20
alveolar macrophage, 21, 58, 74
alveolar type II cells, 29
alveoli, 48, 49, 58
ambroxol, 68
anaphylaxis, 54
angiogenesis, 54
anticoagulant, 44
anticoagulants, viii, 37, 41
antioxidant, 54, 56, 68, 69, 70
APC, 41
apoptosis, 28, 38, 53, 54, 56, 58, 60
ARDS, viii, 2, 20, 21, 37, 38, 39, 40, 41, 42, 45, 46, 47, 48, 49, 51, 52, 53, 54, 55, 57, 58, 59, 60, 61, 62, 63, 66, 67, 68, 71, 73
arterial hypertension, 52, 66
artery, 28, 32, 38, 46, 50
aspiration, viii, 40, 43, 45, 47, 50, 58, 62
atelectasis, 48, 50
ATP, 27

B

barriers, viii, 19, 59
barriers to entry, 59
base, 59
beneficial effect, viii, 37, 41
benefits, 60
bias, 57
Bilateral, 2
bioavailability, 54
biomarkers, 57
biosynthesis, 68
blame, 14
bleeding, 52
blood, vii, 1, 2, 4, 6, 18, 47, 48, 49, 50, 52, 65
blood flow, 48, 49, 50, 65
blood flow velocity changes, 65
blood pressure, 49, 52
blood transfusion(s), 2, 4, 6, 18, 47
body weight, 49
bonds, 54
bone marrow, 72
bradycardia, 49, 50
breakdown, 48
breathing, 39

C

Ca^{2+}, 24, 25, 26, 27, 33
Ca^{2+} signals, 26, 27
calcium, 25, 28, 33
cancer, 34
capillary, viii, 21, 22, 27, 30, 45
carbohydrates, 22
cardiac output, 51
cardiac surgery, 6, 10, 11, 14
cardio-pulmonary bypass, vii, 1, 3, 6, 17
cascades, viii, 20, 21
causality, vii, 2, 4, 5, 8, 11, 15
cDNA, 74, 75
cell biology, 32
cell death, 21, 31
cell division, 23
cell line, 57
cell signaling, 28
cell surface, 25
cerebral blood flow, 49, 50
chemiluminescence, 55
chemokines, 38, 60, 61
chemotaxis, 53, 54
chemotherapeutic agent, 16
chest radiography, 46
children, 50, 51, 52, 63, 66
China, 19
circulation, 59, 60, 65
clinical trials, 38, 47, 60, 61, 66
CO_2, 23
coenzyme, 56, 71
collagen, 28, 33, 38, 54
communication, 24, 27, 35
community, 57, 71
compatibility, 64
complement, 56, 70
complexity, 3
compliance, ix, 45, 46, 48, 50, 53, 65
complications, 39, 48, 49, 50
conductance, 35
conference, 17
consensus, 17, 42, 46, 61
constituents, 48
control group, 40
controlled trials, 39, 48, 52, 71
controversial, ix, 24, 46, 55
coordination, 42, 62
correlations, 23
corticosteroids, viii, ix, 37, 38, 39, 42, 43, 46, 47
creatine, 57
cycling, 23
cysteine, 54, 69
cytokines, vii, viii, 1, 2, 20, 21, 22, 23, 26, 38, 56, 58, 60, 61, 68
cytoskeleton, 24, 25
cytotoxicity, 54

D

deaths, 40, 57

degradation, 59
deposition, 28, 38
diffusion, 21, 22, 48
disaster, 23
diseases, vii, viii, 2, 23, 45, 55, 73
disorder, 24, 53, 59
dispersion, 49
disseminated intravascular coagulation, 40, 43
distress, viii, 38, 42, 44, 45, 46, 62, 63, 64, 65, 68
distribution, 48, 64
DNA, 59
double-blind trial, 54
down-regulation, 61
drugs, 56, 65
dyspnea, 2

E

ECM, 28
edema, viii, 20, 21, 22, 23, 27, 29, 38, 45, 47, 48, 49, 55, 58, 60, 67
elastase inhibitors, viii, 37, 47
electron, 30
electron microscopy, 30
electroporation, 74
endothelial cells, 3, 24, 26, 27, 28, 32, 34, 56, 70
endothelial NO synthase, 60
endothelial permeability, viii, 20, 24, 29, 31
endothelium, vii, viii, 1, 3, 4, 5, 6, 8, 9, 10, 16, 17, 19, 20, 21, 24, 29, 33, 34, 35, 56, 59
endotracheal intubation, 49
environment, 49
enzyme, 60
epidemiology, 18, 20
epithelial cells, 21, 22, 27, 28, 30, 35, 36, 53
epithelium, viii, 19, 20, 21, 27, 30, 59, 68
epitopes, 16
equilibrium, 68
etiology, vii, viii, 1, 2, 3, 4, 6, 10, 15, 20
EU, 45
evidence, viii, 20, 24, 46, 57, 69

evolution, 62, 72
exposure, 28, 72
extracellular matrix, 28
extravasation, 70
exudate, 22

F

far right, 6
fibrinolysis, 41
fibrinolytic, 41
fibroblast proliferation, 38
fibroblasts, 54, 59, 60
fibrosis, 47, 54, 68
filtration, 22
fluid, ix, 2, 4, 21, 22, 30, 45, 48, 49, 53, 55, 60, 67, 73
fluid balance, 21
fluid management, ix, 45
force, 24
formation, 24, 26, 47, 48, 52, 55

G

gas exchange, viii, 19, 20, 21, 48, 63, 65, 66
gene expression, 59, 60
gene therapy, 57, 59, 60, 61, 71, 73, 75
gene transfer, 60, 74, 75
genes, 28, 59, 60
genome, 59
glutathione, 54, 55, 68, 69
graph, 5
growth, 54, 57, 60, 61, 74
growth factor, 57, 60, 61, 74
GTPases, 23, 24, 31, 32, 33

H

hematopoietic stem cells, 72
heme, 56, 60, 70, 74
heme oxygenase, 60, 70, 74
heterogeneity, 34, 56, 57
high risk patients, 10, 38
histones, 31

HM, 35
homeostasis, 27, 28, 72
host, 20, 21, 22, 57, 72, 74
human, 2, 24, 29, 33, 34, 41, 44, 54, 56, 58, 60, 66, 67, 70
human leukocyte antigen, 2
hyaline, 47
hydrogen, 54, 68
hydrogen peroxide, 54, 68
hydroxyl, 68
hypercholesterolemia, 56
hypertension, 32, 46, 52, 66
hypotension, 49, 50, 53, 54, 55, 61
hypoxemia, ix, 45, 46, 47, 50, 51, 67
hypoxia, vii, 1, 3, 28, 32, 38, 49

influenza, 31, 74
influenza virus, 74
inhibition, 29, 30, 36, 43, 48, 53, 55, 62, 71
inhibitor, 40, 41, 43, 44, 75
innate immunity, 35, 73
insertion, 59
integrin, 28
integrity, 21, 24, 26, 32
intensive care unit, 6, 9, 39, 50, 53
interference, 25
intervention, 3, 50
Iran, 66
ischemia, 22, 34
isozymes, 28

I

ideal, 38
identification, 17
IFN, 60, 74
IL-8, 57
image, 50
immune activation, 59
immune modulation, 28
immune response, 20, 22, 57, 59, 73
immune system, 59
immunity, 21, 35, 73
immunomodulatory, 56
improvements, 40, 57
in vitro, 53, 67, 70
in vivo, 32, 53, 56, 67, 70
incidence, 3, 33, 38, 48, 55
individuals, 5, 6
induction, 58
infants, 48, 49, 50, 62, 64, 65, 66
infection, vii, 1, 3, 33, 38, 56, 57
inflammasome, 29
inflammation, ix, 22, 23, 27, 29, 30, 31, 33, 34, 35, 41, 45, 48, 56, 57, 58, 59, 60, 61, 62, 70, 71, 74
inflammatory cell migration, 55
inflammatory cells, viii, 20, 21
inflammatory mediators, 38, 60
inflammatory responses, 61, 75

J

Japan, 37, 40

K

K^+, 23, 29, 73, 74
keratinocyte, 60, 61
kinetics, 64

L

lead, ix, 14, 21, 23, 45
leakage, viii, 2, 4, 21, 26, 45, 58
leucocyte, 70
leukocytes, 59
light, 24
lipid peroxidation, 55
lipids, vii, 1, 2, 59
liver, 26, 34, 57
localization, 58
lovastatin, 56
low risk, 10
lower esophageal sphincter, 33
lung disease, 23, 48
lung function, 27
lymphocytes, 22
lymphoma, 22

M

macrophages, 20, 21, 22, 28, 30, 58, 69, 72, 74
magnitude, 10
majority, 10
malignancy, 6, 9, 10, 11, 14, 16, 17
mammals, 27, 28, 35
man, 69
management, ix, 18, 20, 42, 45
marrow, 72
mass, 23
matrix, 28, 53, 68
matrix metalloproteinase, 53, 68
mechanical properties, 29, 35
mechanical ventilation, 6, 9, 28, 39, 40, 48, 49, 53, 55, 69
meconium, 50, 62
media, 29
median, 51
medical, 65
medication, 67
medicine, 59
mesenchymal stem cells, 72, 75
messengers, 27
meta-analysis, 39, 43, 50, 51, 52, 55, 57, 63, 66, 71
metabolism, 54
metabolites, 23
metalloproteinase, 68
methodology, 49
methylprednisolone, 38, 39, 42
mice, 29, 31, 33, 39, 56, 58, 60, 71, 72, 73, 74, 75
microcirculation, 27
migration, viii, 20, 23, 24, 28, 55, 58
MIP, 58
mitogen, 24
MMP, 53
MMP-9, 53
models, vii, ix, 2, 4, 5, 10, 15, 22, 46, 55, 58, 60, 70
modulus, 28, 35
molecular weight, 29
molecules, 21, 23, 27, 38, 55, 59, 69

monolayer, viii, 20, 27
morbidity, viii, 19, 39, 46, 61
mortality, viii, 19, 20, 29, 37, 38, 39, 40, 41, 42, 43, 46, 48, 50, 51, 52, 53, 55, 57, 58, 61, 66, 71
mortality rate, 38, 40, 41, 43
MR, 63, 64
mRNA, 54
mucus, 54, 55
multipotent, 73
mutations, 33
myocardium, 34
myosin, 24

N

Na^+, 23, 29, 60, 73, 74
necrosis, 22, 23, 28, 73
neonates, 50
neovascularization, 47
Netherlands, 1
neutrophils, vii, viii, 1, 2, 4, 6, 8, 9, 10, 17, 20, 21, 22, 31, 53
nitric oxide, 30, 35, 46, 66, 67, 75
nitric oxide synthase, 30, 35, 75
nitrite, 55
nitrogen, 52
nitrogen dioxide, 52
nucleic acid, 59
nucleus, 59

O

opportunities, 3
organ, 29, 39, 41, 52, 55, 57, 58
organelles, 23
oxidative stress, 28, 30
oxygen, 21, 22, 36, 48, 50, 54

P

pancreatitis, viii, 45, 47
parallel, 4
paralysis, 53

parenchyma, 59
participants, 52
pathogenesis, 3, 17, 18, 20, 23, 28, 38, 39, 43, 54, 59, 62
pathogens, 22, 23
pathology, 47, 55
pathophysiological, 3
pathophysiology, 2, 3, 18, 32, 33, 57, 61
pathways, viii, 20, 23, 24, 25, 26, 27, 31, 48, 54
peptides, vii, 1, 2
perfusion, 65
permeability, viii, 20, 21, 23, 24, 26, 28, 29, 30, 31, 32, 33, 34, 55, 61
permit, 27
peroxidation, 55
peroxide, 54, 68
pH, 27
pharmacological treatment, ix, 45
pharmacotherapy, ix, 45, 47
phenotype, 32
phospholipids, 22, 47
phosphorylation, 24, 25, 27, 56
physiology, 32, 33
pilot study, 67
placebo, 38, 39, 41, 53, 54, 55, 56, 69, 71
plasma levels, 57
plasma membrane, 24, 28, 35
plasma proteins, 48
plasmid, 59, 60
platelet aggregation, 52
platelets, 30, 31
PM, 18, 67, 69
pneumonia, viii, 6, 9, 10, 11, 14, 22, 45, 47, 50, 57, 71, 74
pneumonitis, 31
polarity, 24
polymer(s), 49, 59, 64, 73
polymerization, 24
population, 16
premature infant, 49, 64, 65
preparation, 48
preservation, viii, 20
pressure gradient, 22
preterm infants, 48, 62, 65

prevention, vii, 3, 15, 48, 63, 71
prevention of infection, 71
priming, vii, 1, 3
principles, 42
progenitor cells, 56, 58, 60, 71
prognosis, 2, 42
pro-inflammatory, 38, 53, 54, 58, 68
proliferation, 31, 38
promoter, 59
prostacyclins, 51, 67
prostaglandins, 67
protease inhibitors, 47
protection, viii, 20, 32, 70
protective mechanisms, viii, 20
protein family, 27
proteins, 22, 24, 47, 48
proteolysis, 30
Pseudomonas aeruginosa, 74
pulmonary artery pressure, 50
pulmonary circulation, 59, 60
pulmonary diseases, 73
pulmonary edema, viii, 20, 27, 29, 58, 60, 67
pulmonary hypertension, 32
pulmonary microvascular hyperpermeability, vii
pulmonary microvasculature, viii, 19, 20
pulmonary vascular resistance, 48

R

radicals, 54
radiography, 46
reactions, 30
reactive oxygen, 21, 36, 54
reactivity, 61
receptors, 24, 26, 31, 56, 70
recognition, 28
recovery, 69
redistribution, 48
regeneration, 57, 72
rejection, 74
renal dysfunction, 52
renal medulla, 23, 31
repair, 22, 23, 30, 53, 56, 57, 59, 67, 72

replication, 59, 74
researchers, 23, 26
resistance, 48, 61
resolution, 22, 30, 47, 56, 60, 71
respiratory distress syndrome, viii, 2, 10, 20, 31, 37, 38, 42, 43, 44, 45, 46, 61, 62, 63, 64, 65, 66, 67, 68, 69, 72, 73
respiratory dysfunction, 21
respiratory failure, viii, 19, 20, 42, 49, 50, 51, 63, 64, 66
response, vii, 1, 2, 11, 20, 22, 35, 43, 52, 58, 59, 67, 69, 72, 73
retroviruses, 59
RH, 32, 36, 44, 62, 63, 64, 67
risk(s), vii, viii, 2, 3, 4, 5, 6, 10, 15, 38, 39, 49, 51, 52, 54, 57, 59
risk factors, vii, viii, 2, 3, 4, 5, 6, 15
RNA, 59
rodents, 58

S

safety, 3, 43, 44, 52, 59, 61, 67
saturation, 50
scanning electron microscopy, 30
secretion, 53, 68, 73
sensitivity, 16, 25
sensitization, 26, 33
sepsis, vii, viii, 1, 3, 22, 24, 26, 27, 29, 33, 35, 38, 40, 41, 43, 44, 45, 47, 55, 56, 58, 69, 70, 71, 72
septic shock, 41, 42, 44, 63
serum, 26
shear, 64
shock, 22, 41, 42, 44, 63, 75
showing, 52
side effects, 60, 62
signal transduction, 24
signaling pathway, viii, 20, 23, 24, 25, 26, 27, 29, 36, 54
signalling, 70
signals, 26, 27, 30
skeletal muscle, 53
Slovakia, 45, 75
SPA, 68

species, 21, 27, 36, 54
sphincter, 33
SS, 35, 44, 75
stability, 23, 25
state, 4, 25, 68
statin, 57
stem cells, 36, 58, 66, 72, 73, 75
steroids, 16
stress, 28, 30
stromal cells, 57, 72, 73, 75
structure, 34, 62
subgroups, viii, 37, 41
Sun, 31, 35, 62, 66
suppression, 75
surface area, 49
surface tension, 49
surfactant(s), ix, 21, 22, 45, 46, 47, 48, 49, 50, 51, 54, 62, 63, 64, 65, 66
surfactant administration, 48, 49, 64, 65
survival, 38, 40, 41, 54, 55, 56, 58, 60, 70, 72, 73
survival rate, 40, 55
susceptibility, vii, 2
syndrome, viii, 19, 29, 38, 39, 42, 43, 44, 45, 46, 50, 62, 63, 64, 65, 68
synergistic effect, 61
synthesis, 28, 48, 54

T

tachycardia, 54
target, 56, 59, 70, 73, 75
Task Force, 62
technology, 51
temperature, 30, 49
tension, 49
therapeutic approaches, 57
therapeutic effects, 66
therapeutic targets, 57
therapeutic use, 68
therapy, 37, 41, 42, 44, 46, 49, 50, 51, 52, 54, 57, 58, 59, 60, 61, 62, 63, 64, 65, 66, 68, 71, 73, 75
threshold level, 4, 9
thrombin, 41

thrombocytopenia, 40
thrombosis, 4, 5, 18, 41
tissue, 21, 22, 26, 27, 30, 36, 41, 44
TLR, 23, 56
TLR2, 26
TLR4, 25, 26, 33, 34
TNF, 22, 28, 55, 56, 58, 61
TNF-alpha, 56, 58
TNF-α, 28, 55, 58, 61
TRALI, v, vii, 1, 2, 3, 4, 5, 6, 8, 9, 10, 11, 13, 14, 15, 16, 17, 18
transcription, 23, 54, 56, 59, 75
transduction, 24, 25, 59
transfection, 60, 73
transforming growth factor, 60
transfusion, vii, 1, 2, 3, 4, 5, 6, 7, 9, 10, 11, 13, 14, 15, 16, 17, 18, 22, 47
translation, 59
transport, 21, 60
trauma, vii, viii, 1, 3, 22, 45, 47
treatment, viii, ix, 20, 27, 38, 39, 40, 41, 42, 43, 44, 45, 48, 49, 50, 51, 53, 54, 55, 61, 62, 63, 64, 65, 66, 68, 69, 70, 71
trial, 29, 42, 44, 50, 51, 53, 54, 56, 61, 63, 66, 68, 71
triggers, viii, 20
tumor, 22, 23, 28, 73
tumor necrosis factor, 22, 23, 28, 73
tyrosine, 34

U

UK, 53
United States (USA), 19, 33

V

variables, 46
vascular wall, 32
vasculature, 4
vasodilator, 52
vector, 59, 60, 74
vein, 24, 29, 60
velocity, 50, 65
ventilation, ix, 6, 9, 28, 30, 39, 40, 45, 48, 49, 50, 51, 52, 53, 55, 64, 65, 69
venules, 22
viral gene, 59
viral vectors, 59, 60
viruses, 59
viscosity, 49, 54, 64

W

walking, 11
water, 22, 57
weakness, 54
wealth, 3